The Language of
Mathematics

The Language of Mathematics

How the Teacher's Knowledge of Mathematics Affects Instruction

Edited by Patrick M. Jenlink

ROWMAN & LITTLEFIELD
Lanham • Boulder • New York • London

Published by Rowman & Littlefield
An imprint of The Rowman & Littlefield Publishing Group, Inc.
4501 Forbes Boulevard, Suite 200, Lanham, Maryland 20706
www.rowman.com

6 Tinworth Street, London SE11 5AL, United Kingdom

British Library Cataloguing in Publication Information Available

Library of Congress Cataloging-in-Publication Data Available

ISBN 978-1-4758-5479-4 (cloth : alk. paper)
ISBN 978-1-4758-5480-0 (pbk. : alk. paper)
ISBN 978-1-4758-5481-7 (electronic)

∞™ The paper used in this publication meets the minimum requirements of
American National Standard for Information Sciences—Permanence of Paper
for Printed Library Materials, ANSI/NISO Z39.48-1992.

Contents

Preface

Mathematics is a living subject that seeks to understand patterns that permeate both the world around us, and the mind within us. It is both a language and a form of logical reasoning with which we interact with the world and express our understanding and justify our decisions. Although the language of mathematics is based on rules that must be learned, it is important for motivation that students move beyond rules to be able to express things in the language of mathematics. Also important is for students to understand how to create reasoned, logical arguments for solutions when engaged in problem solving and present a valid explanation using the language of mathematics. Therein lies the challenge for teachers of mathematics.

Mathematics equips students with a uniquely powerful set of tools to understand and change the world. These tools include logical reasoning, problem-solving skills, and the ability to think in abstract ways. Mathematics can be seen as a way of organizing ideas in order to develop concepts, and constructing knowledge is dependent upon concepts. Mathematical learning is associated with the development of mathematical understanding.

The framing of knowledge and understanding for teaching mathematics has centered on the question, What mathematical language, reasoning, insight, understanding, and skills are required for a person to teach mathematics? Teacher educators and other scholars have investigated the nature and effects of teachers' mathematical knowledge for many decades. These studies have drawn attention to substantial mathematical issues that arise in day-to-day school instruction but are not well understood by U.S. pre-service teachers when they graduate from college. International studies also highlight the importance of continuing study as an integral part of a teacher's day-to-day duties.

What we know, in part, is that mastering the language of mathematics is an essential aspect of teaching mathematics to students at all grade levels.

This process continues throughout an individual's mathematics education and well into his or her application of mathematical knowledge and understanding. Thus college mathematics courses should be designed, in concert with teacher preparation courses, to prepare prospective teachers for the lifelong learning of mathematics rather than to teach them all they might need to know in order to teach mathematics well.

Importantly, the mathematical knowledge needed for teaching is quite different from that required by college students pursuing other mathematics-related professions. Pre-service teachers need a solid understanding of mathematics so that they can teach it as a coherent, reasoned activity and communicate its elegance and power. Teacher educators who understand mathematics and have a firm grasp of the language and an effective knowledge of mathematics are particularly qualified to teach mathematics in the connected, sense-making way that students need.

Developing the language and a knowledge of mathematics is an essential aspect of teaching mathematics to students at all grade levels, and this process continues throughout an individual's mathematics education. Because the understanding of mathematical language affords access to concepts and contributes to an understanding of logical reasoning, prospective teachers at all levels need to experience justifying conjectures with informal but valid arguments if they are to make mathematical reasoning a part of their teaching.

The daunting challenge before us now as teacher educators and content area specialists in mathematics is how to achieve the goal of preparing proficient mathematics teachers given the diverse mathematical preparation of college students who will become teachers and changing views about what mathematical knowledge and language is needed to be an effective teacher.

There is without question a need for developing in teachers a "deep understanding" of school mathematics concepts and procedures. The emphasis is on the mathematics that teachers need to know but also there is recognition that teachers must develop "mathematical knowledge for teaching." This knowledge is necessary for teachers to assess their students' work, recognizing both the sources of student errors and their students' understanding of the mathematics being taught. Important is prospective teachers having a solid understanding of the mathematics that they will teach as well as a knowledge of how to teach mathematics.

The Language of Mathematics: How the Teacher's Knowledge of Mathematics Affects Instruction introduces the reader to a collection of thoughtful works that represent current thinking on mathematics teacher preparation. Each chapter focuses on mathematics teaching and the preparation of teachers who will enter classrooms to instruct the next generation of students in mathematics. Chapter 1 opens the book with a focus on the language and knowledge of mathematics teaching, providing the reader with an introduction. The authors of chapters 2–9 present field-based research that examines the complexities of content and pedagogical knowledge as well as knowledge for teaching. Each chapter offers the reader an examination of mathematics teacher preparation and practice based on formal research that will provide the reader with insight into how the research study was conducted and, equally important, the findings and conclusions drawn with respect to mathematics teacher preparation and practice. Finally, chapter 10 presents an epilogue that focuses on the future of mathematics teacher preparation.

Acknowledgments

The idea for this book began as a conversation among colleagues focused on the importance of "mathematical knowledge for teaching" and the preparation of teachers. Equally influential was a focus on mathematics language in relation to mathematics content knowledge and pedagogical content knowledge. The concern we fostered was the need to ensure that the next generation of mathematics teachers is prepared to meet the learning needs of the new generations of youth that enter classrooms. This, we believed, required a focus on research.

In our conversations, we asked what worked and what didn't work and why with respect to teaching mathematics. We recognized that using a subject-specific framework for researching teachers' knowledge could give researchers, teacher educators, and teacher practitioners a window into the ways in which teacher knowledge influences the work that they do with students.

Acknowledgment and thanks go to the contributing authors whose research offers insight and thoughtful considerations for understanding the need to examine the language and knowledge of mathematics as related to teacher preparation and practice and the need for a "mathematical knowledge for teaching."

The authors bring the welcomed perspective of theorists and researchers as well as their own field-based research to the discourses presented in the book. Without the contributing authors, this book would not have been possible. The authors of the chapters examining mathematics and educator preparation bring their considerable experience to bear on interpreting the complexity, challenges, and problems associated with preparing each new generation of mathematics teachers.

Gratitude is extended to the external reviewers who took time out of their busy schedule to review and provide comments and suggestions on the chapters. Acknowledging the value of the chapters and offering constructive feedback was invaluable, as was the affirmation by reviewers of both the need and the importance of a book committed to mathematics teaching and the preparation of both mathematics knowledge and language proficient teachers.

Likewise, gratitude is extended to Tom Koerner and the editorial staff at Rowman & Littlefield for their vision in seeing the value of a book on mathematics teaching that draws into specific relief the need to advance an understanding of how the language and knowledge of mathematics as well as a "mathematical knowledge for teaching" present new challenges for teacher preparation and practice in public school and college classrooms and, most importantly, in educator preparation programs.

As well, thanks are in order for the production staff at Rowman & Littlefield for their ever-vigilant efforts to bring the book to completion. Working with a quality publisher and the folks that do the work to translate a manuscript into a completed book is a rewarding experience.

Finally, gratitude goes to Stephen F. Austin State University for supporting this project and enabling the realization of a work that will shape educator preparation for years to come.

Chapter 1

Understanding the Language and Knowledge of Mathematics

Preparing Mathematics Teachers of Substantive Knowledge

Patrick M. Jenlink

Mathematics is a universal language, a system of symbols and mathematical meaning shared and understood across cultures and disciplinary boundaries. Understanding mathematics requires mathematical knowledge (content knowledge), mathematical reasoning (cognitive reasoning), and symbolic understanding (ability to interpret symbols; see Ball, Hill, & Bass [2005]; Lannin et al. [2013]; and Steele & Rogers [2012]). Individuals represent their reasoning with mathematical language and symbols so that others can understand what they are thinking. The language of mathematics is perhaps one of the most important in terms of communicating and problem solving in an increasingly complex and diverse world. Importantly, mathematical understanding influences decision making in all areas of life—private, social, and civil (Anthony & Walshaw, 2009a).

Mathematics as a symbolic language system holds a primary position in other disciplines such as science, technology, and engineering, disciplines that, along with mathematics, comprise STEM education; mathematics is the primary language across and within the disciplines that enables each discipline to connect and communicate (President's Council of Advisors on Science and Technology, 2010). Mathematics also holds a primary role in innovation and creating new technologies, medical breakthroughs, space exploration, and a myriad assortment of applications that contribute to a new and better future. On this point of a better future, mathematics is equally critical to enhancing life functions and problem solving, creating a globally interdependent society, researching and addressing climate change, innovating manufacturing, creating alternative fuels, advancing space exploration and travel, and much more.

Mathematics education is a key to increasing the post-school and citizenship opportunities of young people. Yet today, as in the past, many students struggle with mathematics and become disaffected as they continually en-counter obstacles to engagement.[1] The National Mathematics Advisory Panel (2008) reported evidence demonstrating that variation in teacher quality could account for a substantial fraction of the total variation (as much as 12 to 14 percent) in mathematics learning by elementary students in a given school year.[2] Zazkis and Leikin (2010) noted significant gaps between secondary school mathematics and tertiary mathematics. "Students, even those identi-fied in school as high-achieving students, experience unexpected difficulties when beginning undergraduate mathematics courses, and many teachers per-ceive their undergraduate studies of mathematics as having little relevance to their teaching practice" (p. 263).

Considering the scope of impact that mathematics has as a universal language and its role as a cross-disciplinary language that affects math-ematical teaching, Copur-Gencturk (2015) posited two questions of import concerning the effect of mathematical knowledge on teaching: "What role does teachers' mathematical knowledge play in their instruction? Which as-pects of instruction are improved when teachers increase their mathematical knowledge?" (p. 280). The key to answering these questions lies in under-standing the nature of teachers' mathematical knowledge in relation to their pedagogical practices. What constitutes mathematics education is not easily addressed in a concise response and yet we know that preparing mathematics teachers with substantive knowledge of mathematics and pedagogy is only part of the complex equation.

As teacher educators preparing mathematics teachers for the demands of teaching mathematics to students of differing cognitive ability and grade level, we are presented with the critical responsibility of understanding what effective mathematics teaching looks like—and what teachers can do to ad-dress the quality of mathematical knowledge necessary to effectively teach students and avert the struggles that elementary and secondary students encounter (Anthony & Walshaw, 2009a, 2009b).[3] This will certainly require an interweaving of mathematical content knowledge and pedagogical knowl-edge in teaching and learning (Ball & Bass, 2000; Ball et al., 2005; Begle, 1979; Carpenter, Fennema, Peterson, & Carey, 1988; Carpenter, Fennema, Peterson, Chiang, & Loef, 1989; Hiebert & Grouws, 2007; Hill, Rowan, & Ball, 2005; Lannin et al., 2013; Martin, 2007; Steele & Rogers, 2012; Wat-son & Geest, 2005; Zazkis & Leikin, 2009). The complexity of the teaching of mathematics is undeniable and all too often ignored in the preparation of teachers with substantive mathematical content knowledge and mathematical teaching knowledge.[4]

The question that teacher education conjoined with mathematics education is confronted with is, What does "mathematics teachers of substantive knowledge" mean? I argue that mathematics teachers must understand the language of mathematics, but this is only a first step. There are necessary steps to consider in addition to that of the language of mathematics.

THE LANGUAGE OF MATHEMATICS

Learning the language of mathematics is a critical step in understanding and expressing mathematical ideas. Teachers help pupils learn mathematical language to express ideas they already have. Many young children as well as adults understand the concept of multiplication as repeated addition of sets, for instance, and yet do not know how to represent multiplication symbolically. Others may intuitively work with proportional relationships but not be able to express their ideas in mathematical terms or symbols. Mathematics as language is primary in guiding individuals' mathematical development by engaging them in problems, facilitating the sharing of their solutions, observing and listening carefully to their ideas and explanations, and discerning and making explicit the mathematical ideas presented in the solutions (Suurtamm & Vezina, 2010).

Ball, Thames, and Phelps (2008) examined the importance of mathematical language, arguing that teachers need extended expertise with certain mathematical practices. They need to be able to talk explicitly about how mathematical language is used (e.g., how the mathematical meaning of "edge" is different from an everyday reference to the edge of a table); how to choose, make, and use mathematical representations effectively (recognizing mathematical advantages and disadvantages for different options); and how to explain and justify one's mathematical ideas (e.g., why you invert and multiply to divide fractions). All of these are examples where the language of mathematics is central to the ways in which teachers engage in particular mathematical practices and work with mathematics in its decompressed or unpacked form.

Whereas the language of mathematics informs and connects disciplines and professions, the mathematical demands of teaching require specialized mathematical knowledge needed by teachers but not needed by other professions in a similar way. The demands of the work of teaching mathematics create the need for a body of mathematical knowledge and a language of mathematics that is specific to teaching. This language of mathematics in relation to teaching is complex, consisting of a language of mathematical knowledge (content), a language of mathematical teaching knowledge (pedagogy), and a language of mathematical reasoning (cognition; see Ball & Bass, 2000; Ball et al., 2005; Ball et al., 2008; Hill, Sleep, Lewis, & Ball, 2007; Lannin et al., 2013).

Understanding the language of mathematics and its relationship to mathematics education, it is argued, should demonstrate both the beauty of mathematics (with opportunities to discover patterns and solve complex problems) and the utility of mathematics and of computational tools and methods in other fields like science, technology, and engineering. Accomplishing this will require teachers who understand mathematics at a deep level, both content and cognition, and have the instructional and pedagogical acumen necessary to converse in mathematics with students at a substantive level (PCAST, 2010, p. 6). The language of mathematics as a symbolic representation of logic and meaning is connected with learning progressions that interpret and make visible to mathematics teachers the hierarchical understandings students obtain in mathematics.[5] An understanding of the language of mathematics is necessary for understanding the concepts that all children must acquire and serves to highlight common difficulties students face that hinder learning (PCAST, 2010). However, the question remains as to what constitutes a mathematics teacher of substantive knowledge.

TEACHERS OF SUBSTANTIVE KNOWLEDGE

The inescapable point argued here is that there is a powerful relationship between what a teacher knows, how he or she knows it, and what he or she can do in the context of instruction—the mathematical knowledge for teaching (MKT) in relation to the contexts within which teaching takes place (Hill et al., 2008; Phelps & Howell, 2016). Certainly the language of mathematics is a factor in this relationship. MKT includes a range of mathematics knowledge used in teaching the subject in a variety of educational contexts. This MKT is applied knowledge that teachers draw on and use as they engage in and carry out the many practices that translate mathematical knowledge for students, and it is the primary work of mathematics teaching (Ball & Bass, 2002; Ball et al., 2008; Hill et al., 2008).

CONTEXT KNOWLEDGE

A key variable in mathematics teaching is the context within which the mathematics teaching takes place. Teaching context serves a variety of functions, some more critical than others (Phelps & Howell, 2016). Simply stated, "context matters" (p. 53).

Three major components of MKT include features of

students such as their history, learning needs, and actions; the *content* and how it is situated in the curriculum of school learning; and, the *setting*, which includes class size or grouping or mode of instruction such as lecture or discussion. Not only are these particular features central to instruction, but they have also recurred in many different heuristics and models used to characterize instruction. (p. 60)[6]

Context knowledge is a critical variable in the equation of teaching mathematics. Knowing the complexity of context as a teacher and knowing the implications of that complexity for the student is a necessary relationship in preparing teachers of substantive knowledge (see Hill, Ball, & Schilling [2008]; and Hill et al. [2008]).

Shulman's (1986, 1987)[7] original arguments about pedagogical content knowledge included knowledge not only for working in different contexts but equally important for working effectively with the multiplicity of discourses students, teacher, curriculum, and school bring into the classroom (see Hauk, Jackson, Nair, & Y Tsay [2014]). The teaching of mathematics requires not only the language of mathematics but also an understanding that each discourse includes a cultural context. Hauk, Jackson, Nair, and Y Tsay argued that "the ways that teachers and learners are aware of and respond to multiple cultures is a consequence of their orientation towards cultural difference, their intercultural orientation" (p. 22).

CONTENT AND STUDENT KNOWLEDGE

Another key variable in the equation of mathematics teaching is knowledge of content and students (KCS).[8] Hill, Ball, and Schilling (2008) explained that content knowledge intertwined with knowledge of how students think about, know, or learn this mathematics content is critical. KCS is "used in tasks of teaching that involve attending to both the specific content and something particular about learners, for instance, how students typically learn to add fractions and the mistakes or misconceptions that commonly arise during this process" (p. 375). Although professional mathematicians will have mathematical knowledge, such as how to produce a definition of even numbers, multiply a two-digit number by a one-digit number, or write five hundred twenty-six as 526, professional mathematicians will not "know the grade levels at which students, on average, master these ideas and tasks. They are also unlikely to be familiar with common errors that students make while developing proficiency with these ideas and procedures" (p. 380). This is a distinction of importance when considering the preparation of mathematics teachers of substantive knowledge.

Hill et al. (2008) advanced the importance of three factors with respect to "how the relationship between mathematical knowledge and the mathematical quality of instruction can be mediated" (p. 499). These three factors included use of curriculum materials, beliefs about mathematics, and the effects of teacher preparation. The curious distinction between mathematicians and mathematics teachers is that of teacher preparation: the quality of teacher preparation programs and the intersection of mathematics content and learning to teach mathematics is a deciding factor of importance. Unfortunately, as Hill et al. (2008) further argued, "each of the three factors identified can either serve to degrade the mathematical quality of instruction or improve it" (p. 499).[9]

PEDAGOGICAL CONTENT KNOWLEDGE

Pedagogical content knowledge is more than an overlap of knowledge that is both pedagogically and mathematically connected. Silverman and Thompson (2008) argue that pedagogical content knowledge is an inextricable blending that is "predicated on coherent and generative understandings of the big mathematical ideas that make up the curriculum" (p. 502). This kind of knowledge and its interrelationship with content knowledge of mathematics is best understood as knowing the ways of representing and formulating the subject matter of mathematics that make it comprehensible to students as well as understanding what makes the learning of specific mathematics knowledge easy or difficult (Even, 1993).

Shulman (1986) in his research on the importance of pedagogical judgments in relation to teaching content knowledge made the argument that

> the syntactic structure of a discipline is the set of ways in which truth or falsehood, validity or invalidity, are established . . . Teachers must not only be capable of defining for students the accepted truths in a domain. They must also be able to explain why a particular proposition is deemed warranted, why it is worth knowing, and how it relates to other propositions, both within the discipline and without, both in theory and in practice . . . This will be important in subsequent pedagogical judgments. (p. 9)

Shulman's argument is as relevant to teaching mathematics as it is to any content knowledge.

Important to note is that Shulman, in his original statements about pedagogical content knowledge, emphasized knowledge for working effectively with the multiplicity of discourses that students, teacher, curriculum, and school bring into the classroom. Each discourse includes a cultural context. Discourses may differ from person to person or group to group (Hauk et al., 2014). Teaching mathematics effectively occurs at the intersection of mathematics language, mathematics content knowledge, and mathematics pedagogical knowledge, all converging to create a space for teaching and learning mathematics.

TEACHING KNOWLEDGE

What constitutes teaching knowledge, Hauk et al. (2014) argued, centers on the question, What mathematical reasoning, insight, understanding, and skills are required for a person to teach mathematics? Many researchers have worked to develop theoretical models and measures to address this question, most notably Ball et al. (2008) and Hill, Ball, and Schilling (2008). Teaching knowledge is, in large part, what teachers need to know and be able to do to effectively carry out the work of teaching mathematics (Ball et al., 2005).

Determining what teachers need to know and be able to do, explicitly framed in terms of the work teachers do, may seem like a minor point, but it is perhaps one of the more significant variables in the teaching of mathematics (Ball et al., 2008). The relationship between mathematical knowledge for teaching and what teachers need to know and be able to do is complex.

An important point in this argument is that we must consider how to decide whether teachers should be taught particular mathematics content by considering when and where such knowledge would have a bearing on what teachers need to do. This suggests a question as to the depth of mathematical knowledge that a mathematics teacher requires in comparison to the mathematical knowledge of a professional mathematician. There is difference here, and the depth of mathematical knowledge differs due to the application of the knowledge. This also suggests that in the case of preparing teachers of mathematics, the connection between content knowledge and teaching knowledge should be made explicit. Defining mathematical knowledge for teaching in this way addresses two important problems: it provides a basis for setting priorities for what teachers are taught, and it increases the likelihood that teachers will be able to use what they are taught when they teach (see Ball & Bass [2003]; Ball et al. [2005]; Ball et al. [2008]; Hill et al. [2005]; Kleickmann et al. [2013]; and Shulman [1986, 1987]).

FINAL REFLECTIONS

The argument presented is one concerned with what a mathematics teacher of substantive knowledge needs to enter the mathematics classroom to teach. Is an advanced knowledge of mathematics such as that required of a mathematician an essential variable in the equation of learning to teach mathematics and then successfully teaching mathematics? The argument has focused on the importance of the language of mathematics and what constitutes teachers of substantive knowledge. Substantive knowledge in this argument is multi-dimensional, including context knowledge, content and student knowledge, and teaching knowledge.

If the mathematical knowledge required for teaching is indeed multidimensional as this argument suggests, then teacher preparation should be organized to help teachers learn the language of mathematics and the range of mathematical knowledge and pedagogical skill they need in focused ways. Toward this goal, teacher preparation programs would require collaboration between mathematics faculty and teacher education faculty. Equally important, the emphasis of teacher preparation must be reconfigured to include a new level of mathematical knowledge and an understanding of the mathematical language essential to successfully preparing teachers to enter classrooms in schools and teach at a new level of quality and commitment.

NOTES

1. In support of this point, the PCAST report stated,

> Schools often lack teachers who know how to teach . . . mathematics effectively, and who know and love their subject well enough to inspire their students. Teachers lack adequate support, including appropriate professional development as well as interesting and intriguing curricula. (p. 4)

2. See PCAST (2010, p. 58) for further discussion. See also Gonzales, Williams, Jocelyn, Roey, Kastberg, and Brenwald (2009); Henningsen and Stein (1997); Hiebert and Grouws (2007); Hill and Ball (2004); Hill et al. (2005); Monk (1994); Wayne and Youngs (2003); and Zazkis and Leikin (2010).

3. Anthony and Walshaw (2009b, p. 149) developed a set of principles based on the recognition that classroom teaching is a complex activity. They argued the classroom learning community is neither static nor linear and is nested within an evolving network involving the school, the wider education system, and the home and local community. They claimed that effective mathematics pedagogy

- acknowledges that all students, irrespective of age, can develop positive mathematical identities and become powerful mathematical learners
- is based on interpersonal respect and sensitivity and is responsive to the multiplicity of cultural heritages, thinking processes, and realities found in everyday classrooms
- is focused on optimizing a range of desirable academic outcomes that include conceptual understanding, procedural fluency, strategic competence, and adaptive reasoning
- is committed to enhancing a range of social outcomes within the mathematics classroom that will contribute to the holistic development of students for productive citizenship

4. Teachers who master mathematics language are able to understand, as Suurtamm and Vezina (2010) explained, that

> students present a variety of ways of thinking about a mathematical problem and teachers may worry whether they will recognize mathematical understanding in all of the representations presented. Although a student may not appear to a teacher to understand a concept, there may actually be sense in their thinking and explanation. When teachers do not attend to student thinking they tend to dismiss what students bring to the mathematical community and instead impose traditional formalized procedures on students. (pp. 1–2)

5. Shulman (1986, 1987) argued that high-quality instruction requires a sophisticated professional knowledge that goes beyond simple rules such as how long to wait for students to respond and includes categories of typologies of professional knowledge essential to substantive teaching. These categories include

- general pedagogical knowledge, with special reference to those broad principles and strategies of classroom management and organization that appear to transcend subject matter
- knowledge of learners and their characteristics
- knowledge of educational contexts, ranging from the workings of the group or classroom and the governance and financing of school districts to the character of communities and cultures
- knowledge of educational ends, purposes, and values and their philosophical and historical grounds
- content knowledge
- curriculum knowledge, with a particular grasp of the materials and programs that serve as "tools of the trade" for teachers
- pedagogical content knowledge, that special amalgam of content and pedagogy that is uniquely the province of teachers, their own special form of professional understanding (Shulman, 1987, p. 8)

These categories were meant to highlight the important role of content knowledge and to situate content-based knowledge in the larger landscape of professional knowledge for teaching. See Ball et al. (2008) for additional discussion.

6. See Andrews (2011); Ball and Bass (2000); Ball et al. (2008); Cohen, Rauden-bush, and Ball (2003); Hill and Lubienski (2007); Hill, Schilling, and Ball (2004); and Hill et al. (2007).

7. Shulman's original work noted that the

> syntactic structure of a discipline is the set of ways in which truth or falsehood, validity or invalidity, are established . . . Teachers must not only be capable of defining for students the accepted truths in a domain. They must also be able to explain why a particular proposition is deemed warranted, why it is worth knowing, and how it relates to other propositions, both within the discipline and without, both in theory and in practice . . . This will be important in subsequent pedagogical judgments. (Shulman, 1986, p. 9)

8. The knowledge of content and student that Hill et al. (2008) discussed is a primary element in Shulman's (1986) pedagogical content knowledge. In Shulman's view, such knowledge is composed of "an understanding of what makes the learning of specific topics easy or difficult: the conceptions and preconceptions that students of different ages and backgrounds bring with them to the learning of those most frequently taught topics and lessons" (p. 9). Shulman noted that research on students' thinking and ideas provides a critical foundation for pedagogical knowledge.

9. Hill et al. (2008) identified four major categories with respect to knowledge of content and student:

- common student errors: identifying and providing explanations for errors, having a sense for what errors arise with what content, etc.
- students' understanding of content: interpreting student productions as sufficient to show understanding, deciding which student productions indicate better understanding, etc.
- student developmental sequences: identifying the problem types, topics, or mathematical activities that are easier or more difficult at particular ages, knowing what students typically learn "first," having a sense for what third graders might be able to do, etc.
- common student computational strategies: being familiar with landmark numbers, fact families, etc. (p. 380)

REFERENCES

Andrews, P. (2011). The cultural location of teachers' mathematical knowledge: Another hidden variable in mathematics education research. In T. Rowland & K. Ruthven (Eds.), *Mathematics knowledge in teaching* (pp. 99–118). New York, NY: Springer.

Anthony, G., & Walshaw, M. (2009a). *Effective pedagogy in mathematics.* Geneva, Switzerland: International Academy of Education and International Bureau of Education.

Anthony, G., & Walshaw, M. (2009b). Characteristics of effective mathematics: A view from the West. *Journal of Mathematics Education, 2*(2), 147–164.

Ball, D., & Bass, H. (2000). Interweaving content and pedagogy in teaching and learning to teach: Knowing and using mathematics. In J. Boaler (Ed.), *Multiple perspectives on the teaching and learning of mathematics* (pp. 83–104). Westport, CT: Ablex.

Ball, D. L., Hill, H. C., & Bass, H. (2005). Knowing mathematics for teaching: Who knows mathematics well enough to teach third grade, and how can we decide? *American Educator, 29*, 12–22.

Ball, D. L., Thames, M. H., & Phelps, G. (2008). Content knowledge for teaching: What makes it special? *Journal of Teacher Education, 59*(5), 389–407.

Begle, E. G. (1979). *Critical variables in mathematics education: Findings from a survey of the empirical literature.* Washington, DC: Mathematical Association of America, National Council of Teachers of Mathematics.

Carpenter, T. P., Fennema, E., Peterson, P. L., & Carey, D. A. (1988). Teachers' pedagogical content knowledge of students' problem solving in elementary arithmetic. *Journal for Research in Mathematics Education, 19*, 29–37.

Carpenter, T. P., Fennema, E., Peterson, P. L., Chiang, C.-P., & Loef, M. (1989). Using knowledge of children's mathematics thinking in classroom teaching: An experimental study. *American Educational Research Journal, 26*, 499–531.

Copur-Gencturk, Y. (2015). The effects of changes in mathematical knowledge on teaching: A longitudinal study of teachers' knowledge and instruction. *Journal for Research in Mathematics Education, 46*(3), 280–330.

Even, R. (1993). Subject-matter knowledge and pedagogical content knowledge: Prospective secondary teachers and the function concept. *Journal for Research in Mathematics Education, 24*(2), 94–116.

Gonzales, P., Williams, T., Jocelyn, L., Roey, S., Kastberg, D., & Brenwald, S. (2009). *Highlights from TIMSS 2007: Mathematics and science achievement of U.S. fourth- and eighth-graders in an international context.* Washington, DC: U.S. Department of Education.

Hauk, S., Toney, A., Jackson, B., Nair, R., & Y Tsay, J-J (2014). Developing a model of pedagogical content knowledge for secondary and post-secondary mathematics. *Dialogic Pedagogy: An International Online Journal, 2*, 16–40. Retrieved from https://dpj.pitt.edu/ojs/index.php/dpj1/index

Henningsen, M., & Stein, M. (1997). Mathematical tasks and student cognition: Classroom-based factors that support and inhibit high-level mathematical thinking and reasoning. *Journal of Research in Mathematics Education, 28*, 524–549.

Hiebert, J., & Grouws, D. A. (2007). The effects of classroom mathematics teaching on students' learning. In F. K. Lester (Ed.), *Second handbook of research on mathematics teaching and learning* (pp. 371–404). Charlotte, NC: Information Age Publishers.

Hill, H. C., & Ball, D. L. (2004). Learning mathematics for teaching: Results from California's Mathematics Professional Development Institutes. *Journal of Research in Mathematics Education, 35*, 330–351.

Hill, H. C., Ball, D. L., & Schilling, S. G. (2008). Unpacking pedagogical content knowledge: Conceptualizing and measuring teachers' topic-specific knowledge of student. *Journal for Research in Mathematics Education, 39*(4), 372–400.

Hill, H. C., Blunk M. L., Charalambous, C. Y., Lewis, J. M., Phelps, G. C., Sleep, L., & Ball, D. L. (2008). Mathematical knowledge for teaching and the mathematical quality of instruction: An exploratory study. *Cognition and Instruction, 26*(4), 430–511.

Hill, H. C., & Lubienski, S. T. (2007). Teachers' mathematics knowledge for teaching and school context: A study of California teachers. *Educational Policy, 21*(5), 747–768.

Hill, H. C., Rowan, B., & Ball, D. L. (2005). Effects of teachers' mathematical knowledge for teaching on student achievement. *American Educational Research Journal, 42*, 371–406.

Hill, H. C., Schilling, S. G., & Ball, D. L. (2004). Developing measures of teachers' mathematics knowledge for teaching. *Elementary School Journal, 105*(1), 11–30.

Hill, H. C., Sleep, L., Lewis, J., & Ball, D. L. (2007). Assessing teachers' mathematical knowledge: What knowledge matters. In F. Lester (Ed.), *Second handbook of research on mathematics teaching and learning* (pp. 111–156). Charlotte, NC: Information Age Publishing.

Kleickmann, T., Richter, D., Kunter, M., Elsner, J., Besser, M., Krauss, S., & Baumert, J. (2013). Teacher's content knowledge and pedagogical content knowledge: The role of structural differences in teacher education. *Journal of Teacher Education, 64*(1), 90–106.

Lannin, J. K., Webb, M., Chval, K., Arbaugh, F., Hicks, S., Taylor, C., & Bruton, R. (2013). The development of beginning mathematics teacher pedagogical content knowledge. *Journal of Mathematics Teacher Education, 16*(6), 403–426.

Martin, T. S. (Ed.). (2007). *Mathematics teaching today: Improving practice, improving student learning* (2nd ed.). Reston, VA: Author.

Monk, D. H. (1994). Subject area preparation of secondary mathematics and science teachers and student achievement. *Economics of Education Review, 13*, 125–145.

National Mathematics Advisory Panel. (2008). *Foundations for success: The final report of the National Mathematics Advisory Panel*. Washington, DC: U.S. Department of Education.

Phelps, G., & Howell, H. (2016). Assessing mathematical knowledge for teaching: The role of teaching context. *Mathematics Enthusiast, 13*(1–2), 52–70.

President's Council of Advisors on Science and Technology. (2010). *Report to the president: Prepare and inspire: K–12 education in science, technology, engineering, and math (STEM) for America's future*. Washington, DC: Executive Office of the President, President's Council of Advisors on Science and Technology.

Shulman, L. S. (1986). Those who understand: Knowledge growth in teaching. *Educational Researcher, 15*(2), 4–14.

Shulman, L. S. (1987). Knowledge and teaching: Foundations of the new reform. *Harvard Educational Review, 57*, 1–22.

Silverman, J., & Thompson, P. W. (2008). Toward a framework for the development of mathematical knowledge for teaching. *Journal of Mathematics Teacher Education, 11*, 499–511.

Steele, M. D., & Rogers, K. C. (2012). Relationships between mathematical knowledge for teaching and teaching practice: The case of proof. *Journal of Math Teacher Education, 15*, 159–180.

Suurtamm, C., & Vezina, N. (2010). Transforming pedagogical practice in mathematics: Moving from telling to listening. *International Journal for Mathematics Teaching and Learning, 31*, 1–19.

Watson, A., & De Geest, E. (2005). Principled teaching for deep progress: Improving mathematical learning beyond methods and material. *Educational Studies in Mathematics, 58*, 209–234.

Wayne, A. J., & Youngs, P. (2003). Teacher characteristics and student achievement gains: A review. *Review of Educational Research, 73*, 89–122.

Zazkis R., & Leikin, R. (2009). *Advanced mathematical knowledge: How is it used in teaching?* Electronic proceedings of the Sixth Conference of the European Society for Research in Mathematics Education (CERME-6) (pp. 2366– 2375). Retrieved from www.inrp.fr/editions/cerme6

Zazkis, R., & Leikin, R. (2010). Advanced mathematical knowledge in teaching practice: Perceptions of secondary mathematics teachers. *Mathematical Thinking and Learning, 12*(4), 263–281.

Chapter 2

Cultivating Dispositions for Teaching and Learning Elementary Mathematics

Michelle C. Hughes

INTRODUCTION

Baseball Hall of Famer Yogi Berra used to say that baseball is 99% a mental battle. These words ring true for learning, understanding, and teaching mathematics. Learning mathematics can be a great challenge. Often when individuals, young and old, are asked about their mathematics experiences in school, strong reactions often result. Misconceptions about mathematics, based on individual experiences, frequently impact students' mathematical learning and perceptions (Hemmings, Grootenboer, & Kay, 2011).

Misconceptions can hinder the learning process for students over the short-term and even the long-term. To shift negative misconceptions, develop positive attitudes, and foster expertise for teaching elementary mathematics, intentional efforts and training are needed beginning in pre-service preparation (Hart, Oesterle, & Swars, 2013; Lin & Tai, 2016).

As an instructor of pre-service teachers in an undergraduate preparation program, every semester I ask pre-service teachers in my mathematics methods course to describe a color or object that reveals their own experience as a learner of mathematics. Some pre-service teachers describe their experience as black: negative, difficult, or frustrating. Some describe their mathematics experience as yellow: cheery, challenging, or even fun to tackle. Each semester I ponder the varied responses and wrestle with the challenge of how to influence and alter attitudes about mathematics. In this chapter, I will explore how pre-service programs integrate and connect dispositional development and mathematics experiences to positively impact teacher attitudes for teaching elementary mathematics.

LITERATURE REVIEW

Pre-service teachers must be prepared with content, pedagogy, and a professional disposition for classroom impact. This case study explores the dispositions and attitudes needed for teaching elementary school mathematics. The review of literature for this study includes three sections: Math and Emotion, Math Attitudes, and Math Confidence.

Math and Emotion

Teachers are under extreme pressure to meet the needs of all students in the mathematics classroom. With increased expectations to improve student achievement and meet the social and emotional needs of students, the responsibility to foster positive attitudes about mathematics instruction as well as to connect mathematics to students' everyday lives is more important than ever.

Willis (2010) suggested that for children to embrace mathematics, they must sense a safe environment for learning. This approach coincided with the introduction of common core state standards (California Department of Education, 2013) that shifted away from the traditional "drill and kill" of mathematics practice, memorization, more practice, and recall. This change required teachers to be boldly thoughtful with mathematics instruction as well as deliberate in creating a positive classroom culture for learning mathematics.

In 2006, Goetz, Frenzel, Pekrun, and Hall concluded that teachers must be aware of students' emotional experiences in classroom instruction. The authors found that letter grades and enjoyment of the subject matter contributed to students' perceptions of content. The researchers recommended creating content-specific, user-friendly learning communities that included students' social and emotional responses to classroom content because when the social, the emotional, and academic learning are interconnected, it can influence student perceptions. Another study of interest promoted mindfulness to decrease teacher and student mathematics anxiety (Hemmings, Grootenboer, & Kay, 2011).

Several years later, Safir (2017) suggested developing listening skills and emotional intelligence to foster empathy and build relationships to transform schools. Recent neuroscience research has also demonstrated meaningful links between attitudes, belief, math practices, and student performance (Schwartz, 2015).

Math Attitudes

One of the greatest roadblocks in mathematics learning is the perception that mathematics is difficult and confusing. When content becomes difficult for students to understand, anxiety often results, and these negative feelings can provoke negative emotional responses. Most often, math anxiety contributes to a general lack of confidence in mathematics learners. Stuart (2000) concluded that teacher attitudes influence students' mathematical confidence and suggested that to feel good about mathematics, students need to believe that they are capable of doing math.

Additional research has suggested that what is effective for some students is not always effective for others (Soto-Johnson, Iiams, Oberg, Boschmans, & Hoffmeister, 2008). Thus, implementing a variety of strategies and techniques in mathematics can decrease anxiety and increase positive attitudes toward mathematics content. This belief has led some in the field to ponder how to build student confidence for learning and understanding mathematics. One suggested strategy is to survey student attitudes about mathematics. This recommendation suggests frequently asking students about their math perceptions and their best and worst experiences with math. Another suggestion recommends removing obstacles such as time limits that can hinder students' math ability and learning (Schwartz, 2015).

Early on, educational reformer John Dewey suggested that reflective thinking is an active process (1916, 1938). Dewey emphasized the importance of cultivating dispositions, such as reflection, to create meaningful experiences in and out of the classroom. Years later, the literature built on this idea and encouraged teachers to establish a climate conducive to student learning, reflection, and classroom success (Jones, Jones, & Vermette, 2009). Some researchers urged pre-service programs to develop reflective habits in candidates to grow awareness of the importance of learning to teaching mathematics conceptually as well as to identify emotions induced from mathematics experiences (Soto-Johnson, Iiams, Oberg, Boschmans, & Hoffmeister, 2008).

Reflections that reinforce conceptual understanding about mathematics, articulate teaching objectives, and confront emotional reactions to mathematics were recommended for pre-service teachers. Greshem (2008) concluded that when pre-service teachers have low math anxiety, they also have high mathematics efficacy, while Hart, Oesterle, and Swars (2013) suggested that when teachers identify awareness about their own math experiences, they can be more intentional in molding and changing students' math attitudes.

Mathematics expert Marilyn Burns has acknowledged the need to deepen students' knowledge and appreciation of mathematics (2007). In order to transform mathematical teaching practices, teachers should examine student attitudes, past and present, and student emotions about mathematics. When teachers foster optimism toward math, students engage with the material and have increased motivation (Willis, 2008). For educators to influence and shift student attitudes about mathematics, they must encourage students to value mathematics with the end goal of developing a lifelong appreciation for mathematics and learning (Royster, Harris, & Schoeps, 1999).

Math Confidence

Additional recommendations from the literature encourage mathematics teachers to serve in a coach or facilitator role to build self-confidence and refine students' math skills. Stuart and Thurlow (2000) proposed that teachers tackle math anxiety with a coach-like attitude; specifically, if teachers share passion and excitement for mathematics, their enthusiasm becomes contagious and transfers to students. When enthusiasm spreads, students are more willing to take risks. Hence, when a classroom climate is positive and affirming of math, positive attitudes for elementary mathematicians are reinforced (Willis, 2010).

The reviewed literature reveals the need for a supportive environment for nurturing mathematical learning and establishing a community of mathematical thinkers. Practice strategies, positive classroom messaging for teamwork, perseverance, and risk-taking are highlighted (Bray, 2014). Some researchers encourage the use of strategic vocabulary to invite numerous solutions and strategies for problem solving. For example, when educators employ even a simple teaching technique like wait time, students recognize the space and freedom to strategize.

Furthermore, when students make mistakes in the mathematics classroom, they need to know that their mistakes are part of the learning process. Showing students common errors and helping them examine problems to discover multiple strategies alters misconceptions about math (Willis, 2010). Thus, understanding the value of making mistakes and learning from mistakes emerges as a strong priority for the elementary mathematics classroom (Schwartz, 2015).

The literature highlights associations with Dweck's growth mindset theory that endorses changing perceptions of learning (2006, 2016). Dweck's focus on the learning process, student efficacy, and student determination can inform and mold classroom attitudes. Additionally, this focus on habits of mind, or attributes, like resilience and grit, affirms earlier research that suggested student and teacher perspectives can be altered to improve math performance (Bray, 2015; Lin & Tai, 2016; Oaklander, 2015).

RESEARCH QUESTIONS

The reviewed literature affirms the significance and benefit of cultivating dispositions and attitudes in new teachers of elementary mathematics, yet the question of how to foster and develop dispositions in teacher candidates prompted the researcher to investigate the impact of dispositions on the developing candidate learning to teach elementary mathematics in pre-service and in their first elementary teaching position. Understanding that learning to teach mathematics is challenging and multifaceted, the researcher examined the following questions:

1. What dispositions contribute to positive or changed attitudes for teaching elementary mathematics?
2. How do pre-service programs promote positive attitudes and dispositions for teaching elementary mathematics?
3. What role do dispositions play in shaping teacher attitudes for teaching elementary mathematics in and beyond a pre-service program?

For the purposes of this study, *dispositions* are defined as the habits of professional action and moral commitments that underlie an educator's performance (Council for the Accreditation of Educator Preparation Standards, 2016). Additional terms used in the study are (1) *dispositional awareness*, or the conscious perception or self-awareness to name, define, and understand professional teaching dispositions; and (2) *dispositional development*, or the ongoing process of cultivating and applying professional teaching dispositions in practice.

METHODOLOGY

The nature of qualitative research is to understand the personal experiences of individuals (Creswell, 2013). The researcher welcomed the opportunity to explore the role of professional teaching dispositions for teaching elementary mathematics. The researcher hoped to advance the program of study as well as contribute to the larger field of pre-service preparation and elementary mathematics instruction. Using a case study framework (Creswell, 2013), the researcher anticipated that the study's data would generate increased understanding of the elementary teacher's dispositional awareness and development specific to learning to teach mathematics in the program of study.

The researcher expected the data to add rich details to the elementary mathematics teacher's profile from pre-service to the first teaching position. A small program of study offers prospective teachers several pathways to earning a bachelor's degree in elementary education as well as a state preliminary multiple-subject credential or state preliminary single credential within four or five years. Students apply to attend the ten-month pre-service program each February. Institutional Review Board approvals are obtained before the start of the study.

In 2016, in an effort to explore the research questions, the researcher emailed electronic surveys to 46 graduates who had earned a preliminary multiple-subject credential from the pre-service program of study between 2012 and 2016 (see Table 2.1).

Table 2.1. Survey Participants

Year in Program	Participants N = 10	Male = M or Female = F	Teaching Assignment, Public or Private School
2012	1	F	7–8th grade, private
2013	2	F	2nd grade, public
			5th grade, private
2014	1	F	4–5th grade, private
2015	5	4 F	3rd grade, public
			2nd grade, private
			6th grade, public
			3rd grade, public
		1 M	1st grade, public
2016	1	F	4th grade, public

Surveys consisted of 18 open-ended, innocuous questions (see Appendix A). An invitation to participate in the study and letter of consent were also emailed to participants. Twenty-two percent of the requested participants completed the surveys, and responses were lengthy. Ethical protocols were followed and detailed field notes were kept. The researcher collected data from former pre-service students rather than current students to minimize risk and potential pressure. Bracketing was used to keep personal bias and assumptions to a minimum. Participants were given the option to skip questions or opt out of the study. Participants represented 10 program graduates teaching in first- through eighth-grade classrooms in California.

Archived artifacts—specifically, end-of-course essays—were collected from the program's mathematics methods course. The end-of-course essay assignment served as a culminating reflection about each pre-service teacher's attitudes and experiences with mathematics during the mathematics methods course. Artifacts were kept confidential and survey responses were identified alphabetically (Participant A–Participant J) to preserve confidentiality and anonymity.

The researcher examined data from the surveys and archived artifacts to capture and accurately describe participants' perceptions and experiences. Survey data was transcribed verbatim, and member checking was used to ensure accuracy. Data was coded and analyzed using a three-step process looking for repetition, categories, patterns, and a priori themes.

The size of the study could be considered a limitation yet the researcher trusted the data to reveal particulars significant to teaching, learning, and thinking about elementary mathematics. Although the sample size was small and could limit generalizations, the researcher predicted results would inform similar-sized programs and advance a larger conversation about dispositions and teaching elementary mathematics. Self-report findings and participant perceptions can be limiting, yet the researcher expected to uncover specifics for dispositional development for teaching elementary mathematics. The direction of the research evolved as understanding of the topic grew.

RESULTS

Four distinct themes and recommendations emerged from data analysis: (1) dispositions are essential for teaching elementary mathematics; (2) developing teachers of elementary mathematics need to become self-aware and understand personal attitudes about mathematics; (3) developing teachers of elementary mathematics need to create a positive, safe space where students can learn mathematics and make mistakes; and (4) developing teachers of mathematics need to link instruction and learning to real-world, relevant connections using a variety of mathematical practices.

DISPOSITIONS ARE ESSENTIAL FOR TEACHING ELEMENTARY MATHEMATICS

Survey data revealed that participant responses affirmed dispositions as an essential component of elementary mathematics instruction (Lin & Tai, 2016; Weis & Herbst, 2015). Participant E defined dispositions as "traits that shape how you come across, how you relate to others, and how you view your profession." Participant G described dispositions in practice as "always learning, serving others, compassionate, always reflecting."

When asked to define professional teaching dispositions, Participant H replied, "I would say that a professional teacher should be humble, steadfast, fun, engaging, creative, a life-long learner, optimistic, dependable, and understanding." Participant I suggested that dispositions are "the values that guide an educator." Additionally, Participant J noted, "Dispositions did play a role in how I learned to teach math. Each of the dispositions guided me in understanding how I could positively teach the subject of math as well as how I could find ways to make the content engaging for students, and support the different needs of my students in the learning of math." Participant J also said, "The [program's] dispositions provided me with points that I could continually reference to make my teaching of math more meaningful to my students."

Archived essays also affirmed the role of dispositions in teaching elementary mathematics: "When thinking about being a compassionate professional in the classroom, I imagine creating a safe place and modeling an attitude towards math that my students may be able to pick up and follow." Additional essay evidence noted, "Being a teacher who portrays compassion may look like understanding the pains and negative emotions towards math but also helping students change their views of this subject."

Another essay said, "I hope that I can show students in my future classroom the joy of patterns and consistency." Yet another essay said, "It was empowering to teach a math lesson because prior to this class [mathematics methods] my perception of math was scary; it [the course] helped me with both my confidence and my feelings towards math."

TEACHERS MUST BE AWARE OF THEIR MATH ATTITUDES

An additional theme noted in the study affirmed the need for teachers, those in pre-service and new to the field, to understand their own attitudes and experiences with and about mathematics (Oesterle & Liljedahl, 2009; Poyraz & Gülten, 2014). Of particular interest, Participant C described professional teaching dispositions as

> a variety of techniques and strategies educators learn, use, and implement that help us succeed in teaching, meaning it helps our students succeed in learning. These dispositions, traits or qualities, help refine the teaching practice so that we are mindful of how we teach, who our students are, and how to best serve a diverse classroom setting.

When surveyed about dispositions specific to mathematics instruction, Participant C stated, "Being a lifelong learner helps me pursue new ideas to try and implement while fostering the same love for learning in my students. Being a compassionate professional allows me to be understanding when students are lost, distraught or discouraged." When asked whether or not dispositions play a role in learning to teach mathematics, Participant C responded, "Yes, an early mindset of putting each disposition into practice [in the pre-service program] with each lesson." This participant also noted that "the ability to reflect after everything has made a huge difference in helping me grow and find new ways to help my students and enhance their learning."

Participant J affirmed this perspective: "A life-long learner is someone who sees mistakes as opportunities to further improve their learning." She also noted, "Being a reflective practitioner also is important when teaching math because when you reflect on an assessment or a lesson you are able to identify the students who might be struggling with a certain concept or who need to be challenged and adjust the next lesson accordingly." Participant J responded,

> The disposition of being a lifelong learner can influence a student's attitude toward math; because if they are not afraid to make mistakes then they won't be afraid to ask questions or ask for help when they need it. If the idea that it is okay to learn from mistakes is continuously reinforced then the student's attitude toward challenging problems will become more flexible and their attitude toward math will hopefully take on a more positive tone.

Artifacts reinforced the survey responses related to self-awareness. One essay said, "I hope to grow as an educator mirroring the dispositions: to always strive to learn more, to be a compassionate professional, and to reflect in order to become a better educator." Another essay observed that "[these] times for reflection have helped me better understand myself and see my goals more clearly." An additional essay said, "My cooperating teacher had a perspective on math that was both a growth mindset, and a desire to tackle math anxiety."

CULTIVATE A POSITIVE, SAFE SPACE IN WHICH TO LEARN MATHEMATICS

Data analysis acknowledged the need to create a positive, safe space where students can learn elementary mathematics. When asked to identify dispositions that influence students, Participant B responded, "We have a no-whining zoning in the classroom, so when we go to do math, which I think is so exciting, we always are thankful rather than bummed at having to work. It really works." This participant highlighted reflection and gratitude as dispositions needed for elementary mathematics teaching. Participant B noted a professional desire to grow in self-awareness: "Because I know that I am capable, given the important role as a math educator, [I am] using the dispositions to grow in my capacity to be the best I can be for my students and their learning." This participant elaborated, "I am excited to teach math as it is my favorite subject."

Likewise, Participant C responded, "A lot of this [teaching mathematics] has been trial and error—practicing what it looks like and seeing what is most effective for student learning." Participant C gave a student example:

> I have a student this year who has had a lot of emotional breakdowns that lead into rage. I have noted that this most frequently happens during math. The [disposition of a] compassionate professional has aided me in each of these breakdowns . . . Compassion has helped [me] understand where he is coming from and what his triggers are as well as how to shower him in love where he needs it most.

Participant I's survey response included, "With students who struggle with math, especially, I try to work alongside them with patience and empathy. With my high achieving math students, I work to challenge them and provide projects that encourage them to take risks as mathematicians."

Participant E built on this idea: "Yes it [dispositions] help create a class-room that is comfortable for students. They feel safe to take risks in their learning because they know I am working to meet them where they are at." This participant continued, "It [dispositions] shape my attitude, which rubs off on my teaching and creates engaged instead of passive learners." Partici-pant E also noted that dispositions help frame her approach to student learn-ing: "I am a learner with them [students], so I remind myself to show patience and compassion as I teach. I reflect daily on their progress, seeing what they need to succeed." She continued, "As a compassionate professional, I [had to] approach [the] students with a gentle and motivating demeanor. I needed to encourage them to struggle to allow themselves to make mistakes, while also holding them accountable to learn the material presented to them." She also noted, "It's been a journey trying to figure out how to best reach every student through small groups, partner work."

When asked to share examples of dispositions used in teaching mathemat-ics, Participant G responded, "My class reflects on our learning constantly. Especially when we get a problem wrong, we go back and reflect on our mis-takes so we don't make them again." This participant elaborated, "We have a no negativity saying in my room and being compassionate and life-long learners helps us remember that we will all learn; it just might take some of us more time and patience." Additionally, Participant G said, "I want my stu-dents to have low math anxiety and feel free to think how they naturally do."

When asked to name dispositions important to teaching elementary math-ematics, Participant E said, "Having a growth mindset, having a positive view on math, believing that everyone is capable of being a mathematician, and being a reflective learner." This participant also said, "Celebrating mistakes during math time shows growth and challenges as a positive mindset." She explained, "They [the students] get daily practice on what it means to face challenges with a positive and capable mind in math. When they reflect on their mistakes and difficulties they are much more likely to persevere and find a way to overcome these mistakes and challenges in math." She continued, "Most of my students love challenging problems during math time. Other students remind struggling students that it is alright to be in the dip because once you reflect, you are able to succeed and grow your mind."

Participant H noted, "Math should always be positive and fun, as well as engaging." Additionally, Participant H stated, "If you are negative about mathematics and do not have a disposition of joy of learning and curiosity to learn more, then students will share your attitude." Participant J said, "The disposition of being a life-long learner is important for teaching math because when you model for your students that it is ok to make mistakes, then, when they come to challenging problems, the students will be more resilient when completing the problem."

Artifacts added reinforcement to the need for a safe space and classroom culture for learning mathematics (Hoerr, 2013; Battey, Neal, & Hunsdon, 2018). One essay revealed, "I can more confidently say that I too can be a compassionate professional when it comes to understanding and respecting student attitudes towards math, but also opening up a window into a new exciting perspective toward math." The author also said, "I will include choice, math menus, and room for creativity when it comes to developing our math minds." Another wrote, "Students will need time to process and experience a learning goal in multiple ways before fully understanding."

LINK MATHEMATICS TO THE REAL WORLD

The theme of real-world connections in mathematics (Burns, 2004, 2007; Suh, 2007) emerged from data analysis. Participant E noted that "math [is] as a tangible thing, math in the real world surrounds us everywhere." Similarly, Participant H said, "This morning, I started my math lesson with a real-world question that no one could solve until we had completed our math lesson. They [the students] became curious and determined to solve this conundrum and ultimately were successful."

Participant C noted,

> I have been trying to share my love for math with the whole class, sharing examples of personal experience and creating real world connections. I talk about what I am still learning and how I love to see them partake in this learning

adventure with me. I have seen multiple girls in my class who once hated math or act afraid of it, start taking more independence in their work, completing problems with more confidence and joy—displaying a love for learning too.

Linking real-world application to a toolbox of mathematical practices surfaced in data analysis. Participant C said,

> I started practicing more gratitude—even for my students. I started being grateful for the little victories and the small things that would happen in math. I started thinking in a more grateful mindset to be able to teach them math and serve them daily. Soon enough, I felt more exited to teach math again and I saw some of my own students who once hated math foster a joy for it too.

When asked whether or not she felt supported in her current teaching position, Participant C responded, "I felt like I had so many wonderful ideas for activities, games, and ways to implement manipulatives in a variety of lessons. I can confidently brainstorm how to engage students with learning and draw their interest levels in during mathematics."

When asked how a new teacher demonstrates dispositions while teaching mathematics, Participant D responded, "Using stories to teach math concepts; stories always engage students and [I am] able to relate to my students with a new math concept. Also having the patience to understand that students learn in a variety of ways and that some students may need more time to understand a concept. I've observed how stories and patience can help my students gain confidence and interest in math." Participant D noted, "I have a great teaching partner. We share ideas and collaborate together."

Participant E described her experience as a new teacher of math: "I entered the field at a time where I need to teach myself strategies for all operations other than the standard algorithm, among other things . . . I need to teach the area model, partial products, etc., so that my students can access content in different ways and think flexibly."

Additionally, Participant H stated that journaling became a helpful tool used in her mathematics classroom: "[I ask students to] journal about how they feel about math. I ask them about what they are unsure of, what they need better explained, what they would like from me, what they feel confident in, and what steps will they take next." Specific dispositions used in math instruction were also identified:

> When I approach students who struggle in math I try to be patient and encourage them that with practice and putting forth their best efforts, they can find success in math. I want my student to feel supported and safe in their learning environment. I want to build their confidence in the subject because students

can become easily frustrated with math at an early age and become unmotivated. (Participant I)

The same participant reflected on experiences with an anxious student: "After I introduced a new concept to the class, I would check in with her when the rest of the class was doing independent work. This would help her stay calm and have more of an open mind about new concepts." Participant I also said, "I try to differentiate instruction to support the variety of learning styles in my class." When asked about whether or not dispositions play a role in learning to teach math, Participant I elaborated, "Students usually have very strong opinions about the subject—either hate it or love it—and I've learned through those dispositions that building their confidence, instilling a lifelong learner mentality, and having them [the students] reflect on their thinking is really helpful."

Furthermore, Participant I suggested, "incorporating some practice building students' math vocabulary" is important in teacher preparation. Participant A said, "How I treat my students impacts their actions, reactions, and attitudes towards and in the classroom."

Several end-of-course essays reinforced the need for real-world connections in the elementary mathematics classroom. One essay said, "I observed one teacher who saved a spot on the whiteboard for a daily mathematical question or fact. This is a great place to bring in everyday mathematical items and objects for students to observe and practice." Another essay noted, "Math should be relevant. Students practice real-world application of mathematics and it helps them make sense of the world."

DISCUSSION, RECOMMENDATIONS, AND IMPLICATIONS

This study affirmed the significance of and need for dispositions in learning to teach elementary mathematics. The researcher noted that over one-half of the study's archived essays explicitly identified a neutral to negative view of math when participants started the pre-service program. Yet when pre-service teachers were given time and practice opportunities with elementary students, they began to acknowledge and develop an appreciation for a supportive classroom culture for teaching math.

Participants observed that sharing a passion for mathematics and feeling passionate about math instruction were privileges. Additionally, participants affirmed the necessity to create a safe environment for making mistakes and taking risks while teaching and learning mathematics. This finding validates research that suggested giving students some control in mathematics to empower them (Schwartz, 2015; Stuart & Thurlow, 2000). When students develop a sense of ownership, they are energized and their confidence increases. Students need opportunities to gain and, when needed, regain control in learning.

One author suggested allowing students to revisit mathematics concepts for mastery to gain a greater feeling of control (Willis, 2010). Willis affirmed that showing students common errors and helping them examine sample problems with multiple strategies can change misconceptions about math. Thus new teacher perceptions about understanding the value of making mistakes and learning from mistakes require intention and the attention of the elementary mathematics teacher. Furthermore, pre-service teachers need to grow self-awareness and take risks when learning to teach mathematics (Nolan, 2011).

Participants also said that real-world connections are necessary for elementary students to make math connections. With the introduction of common core standards and an increased emphasis on practical, real-world application in mathematics, the expectation that students would persevere, strategize, and make relevant connections when solving mathematics problems became a prerequisite for teaching and learning mathematics. Earlier research reinforced fostering a supportive classroom community that encouraged mathematical learning, thinking, problem solving, and discourse, reiterating that students need opportunities to explore and extend their mathematical reasoning (Whitin & Whitin, 2008).

Currently, as the common core's eight mathematical practices (2010) are executed in California's schools, practices such as encouraging the use of academic vocabulary for problem solving, with terms such as *persevere*, *justify*, and *strategize*, are expected. This academic language urges teachers and students to stretch their brains as mathematicians and problem solvers. Additionally, with common core state standards, elementary mathematics teachers are expected to demonstrate and promote the use of multiple strategies to solve mathematics problems. When teachers encourage the use of multiple strategies, students are challenged to explore concepts to build mathematical meaning (Suh, 2007; Van de Walle, Karp, & Bay-Williams, 2013).

Of particular significance, the study revealed a variety of intentional practices that participants implemented to foster positive mathematical attitudes in their classrooms. Participants affirmed that to teach mathematics, self-awareness is critical. To be an ambassador for mathematics, teachers must own their own attitudes and experiences with mathematics. This perception aligns with providing a variety of contexts and activities for new teachers to increase their own dispositional awareness (Freese, 2005) and validates earlier research that encouraged reflection and action (Francis, 1995; Johnson, Vare, & Evers, 2013). Pre-service teachers can develop self-awareness with clear, explicit, and reflective tasks.

Recently, Korthagen, Younghee, and Green (2013) suggested that teachers need time and space to reflect and develop dispositions using the practice of core reflection. Core reflection links reflection, perceptions, attitudes, and abilities to behavior. Teachers must develop classroom attitudes and dispositions to demonstrate that they value math and students' ability to do math.

CONCLUSIONS

This study's findings should challenge pre-service programs, as well as new teachers in their first elementary teaching position, to ponder the importance of nurturing both teacher and student confidence for learning and doing mathematics. Specific, intentional efforts to foster positive attitudes toward mathematics are needed by both teachers and students. This study validates the significance of cultivating dispositions in pre-service preparation for long-term professional practice (Claxton, Costa, & Kallick, 2016; Hart, Oesterle, & Swars, 2013; Oesterle & Liljedahl, 2009).

This study also affirms that dispositions or habits of mind for elementary mathematics can be cultivated (Oaklander, 2015). The results leave room for future exploration of distinct practices that explicitly foster dispositions and dispositional awareness in pre-service preparation and in new teachers. Explicit consideration and examination of how to extend dispositional development for teaching elementary mathematics into the first years of teaching and beyond are recommended for educational leaders and schools. Ongoing training opportunities for dispositional development specific to mathematics can be explored in pre-service programs and schools.

Educators should not assume that all new elementary teachers exit pre-service programs possessing and maintaining dispositional practices and awareness for elementary mathematics. Hence, educators can explore strategic professional development opportunities to nurture dispositions (Costa & Kallick, 2014; Hoerr, 2013) such as perseverance, grit, and efficacy that are integral to teaching elementary mathematics.

Since we live in a time when many kids don't believe they are good at math (Schwarz, 2015), there is an increasing need for pre-service teachers to develop dispositions that model and affirm the struggle, beauty, and joy found in mathematics (Battey et al., 2018; Ograin, 2014). It is imperative that pre-service programs and beginning teachers in their first position infuse elementary classrooms with dispositions for impact with the next generation of mathematicians.

APPENDIX A

Survey Questions

1. What year did you earn your teaching credential?
2. How long have you been teaching?
3. Are you currently teaching in public or private school?
4. What grade level(s)?
5. Please name previous positions and/or grade levels.
6. In your own words, define professional teaching dispositions.
7. Do dispositions inform how you approach and teach K–6 mathematics? Why or why not?
8. In your opinion, name the dispositions that you feel are important to teaching K–6 mathematics.
9. Please name and describe specific examples/times when you demonstrated each of these dispositions teaching math.
10. In your opinion, do the dispositions you named influence students? If so, please give a specific example.
11. In your opinion, do these dispositions influence student attitudes toward mathematics? Please explain.
12. In your opinion, do these dispositions influence your attitude toward teaching mathematics? Please explain.
13. Describe the preparation you received to teach mathematics in your credential program.
14. In your credential program, did dispositions play a role in how you learned to teach mathematics? Why or why not?
15. If dispositions played a role in how you learned to teach mathematics, please give an example(s) from the program.
16. Do you have suggestions for how the credential program can enhance preparation for the teaching of K–6 mathematics?
17. Do you feel supported as a teacher of mathematics in your current teaching position? Please explain.
18. Do you have anything else to share or add?

REFERENCES

Battey, D., Neal, R. A., & Hunsdon, J. (2018). Strategies for caring mathematical interactions. *Teaching Children Mathematics, 24*(7), 432–440.

Bray, W. (2014). Fostering perseverance: Inspiring students to be "doers of hard things." *Teaching Children Mathematics, 21*(1), 5–6.

Burns, M. (2004). 10 big math ideas. *Instructor Magazine*, pp. 16–20.

Burns, M. (2007). *About teaching mathematics: A K–8 resource.* Sausalito, CA: Math Solutions.

California Department of Education. (2013). Common Core State Standards for Mathematics and Standards for Mathematical Practice. Retrieved from www.cde .ca.gov/be/st/ss/documents/ccssmathstandardaug2013.pdf

Claxton, G., Costa, A. L., & Kallick, B. (2016). Hard thinking about soft skills. *Educational Leadership, 73*(6), 60–64.

Costa, A. L., & Kallick, B. (2014). *Dispositions: Reframing teaching and learning.* Thousand Oaks, CA: Corwin.

Council for the Accreditation of Educator Preparation Standards. (2016). http:// caepnet.org/about/vision-mission-goals

Creswell, J. W. (2013). *Qualitative inquiry and research design: Choosing among five approaches.* Thousand Oaks, CA: Sage Publications.

Dewey, J. (1916). *Democracy and education.* New York, NY: Free Press.

Dewey, J. (1938). *Experience and education.* New York, NY: Touchstone.

Dweck, C. S. (2006). *Mindset, the new psychology of success.* New York, NY: Random House.

Dweck, C. (2016). Recognizing and overcoming false growth mindset. *Edutopia.* Retrieved from www.edutopia.org/blog/recognizng-overcoming-false-growth-mindset -carole-dweck

Francis, D. (1995). The reflective journal: A window to preservice teachers' practical knowledge. *Teaching and Teacher Education, 11*(3), 229–241.

Freese, A. (2005). Reframing one's teaching: Discovering our teacher selves through reflection and inquiry. *Teaching and Teacher Education, 22*(1), 100–119.

Goetz, T., Frenzel, A. C., Pekrun, R., & Hall, N. C. (2006). The domain specificity of academic emotional experiences. *Journal of Experimental Education, 75*(1), 5–29.

Greshem, G. (2008). Mathematics anxiety and mathematics teacher efficacy in elementary pre-service teachers. *Teaching Education, 19*(3), 171–184.

Hart, L. C., Oesterle, S., & Swars, S. L. (2013). The juxtaposition of instructor and student perspectives on mathematics courses for elementary teachers. *Educational Studies of Math, 83*(3), 429–451.

Hemmings, B., Grootenboer, P., & Kay, R. (2011). Predicting mathematics achievement: The influence of prior achievement and attitudes. *International Journal of Science and Mathematics Education, 9*(3), 691–705.

Hoerr, T. R. (2013). *Fostering grit.* Alexandria, VA: ASCD.

Johnson, L. E., Vare, J. W., & Evers, R. B. (2013). *Let the theory be your guide: Assessing the moral work of teaching.* In Sanger, M. N. & Osguthorpe, R. D. (Eds.), *The moral work of teaching* (pp. 92–112). New York, NY: Teachers College Press.

Jones, J. L., Jones, K. A., & Vermette, P. J. (2009). Using social and emotional learning to foster academic achievement in secondary mathematics. *American Secondary Education, 37*(3), 4–9.

Korthagen, F. A. J., Younghee, M. K., & Green, W. L. (2013). *Teaching and learning from within: A core reflection approach to quality and inspiration in education.* New York, NY: Routledge.

Lin, S., & Tai, W. (2016). A longitudinal study for types and changes of students' mathematical disposition. *Universal Journal of Educational Research, 4*(8), 1903–1911.

Nolan, K. (2011). Dispositions in the field: Viewing mathematics teacher education through the lens of Bourdieu's social field theory. *Educational Studies in Math, 80*(1), 201–215.

Oaklander, M. (2015). Mindfulness exercises improve kids' math scores. *Time.* Retrieved from http://time.com/3682311/mindfulness-math/

Oesterle, S., & Liljedahl, P. (2009). *Who teaches math for teachers?* Proceedings of the 31st Annual Meeting of the North American Chapter of the International Group for the Psychology of Mathematics Education. Atlanta, GA: Georgia State University.

Ograin, C. (2014, November). *Dispositions of well-prepared students toward mathematics.* Presented at the Santa Barbara Unified School District Gifted and Talented Parent Night for Common Core Math, San Marcos High School, Santa Barbara, California.

Poyraz, C., & Gülten, D. C. (2014). Pre-service mathematics teachers' attitudes towards the profession of teaching. *International Online Journal of Education Sciences, 6*(3), 558–569.

Royster, D. C., Harris, M. K., & Schoeps, N. (1999). Dispositions of college mathematics students. *International Journal of Mathematical Education in Science and Technology, 30*(3), 317–333.

Safir, S. (2017). Learning to listen. *Educational Leadership, 74*(8), 16–21.

Schwartz, K. (2015). *Not a math person: How to remove obstacles to learning math.* KQED. ww2.kqed.org/mindshift/2015/11/30/not-a-math-person

Soto-Johnson, H., Iiams, M., Oberg, T., Boschmans, B., & Hoffmeister, A. (2008). Promoting preservice elementary teachers' awareness of learning and teaching mathematics conceptually through "KTEM." *School Science and Mathematics, 108*(8), 345–54.

Stuart, C., & Thurlow, D. (2000). Making it their own: Preservice teachers' experiences, beliefs, and classroom practices. *Journal of Teacher Education, 51*(2), 113–121.

Suh, J. M. (2007). Tying it all together: Classroom practices that promote mathematical proficiency for all students. *Teaching Children Mathematics, 14*(3), 163–169.

Van de Walle, J. A., Karp, K. S., & Bay-Williams, J. (2013). *Elementary and middle school mathematics: Teaching developmentally* (9th ed.). Upper Saddle River, NJ: Pearson.

Whitin, P., & Whitin, D. J. (2008). Learning to solve problems in primary grades. *Teaching Children Mathematics, 14*(7), 426–432.

Willis, J. (2010). *Learning to love math: Teaching strategies that change student attitudes and get results.* Alexandria, VA: ASCD.

Chapter 3

Pre-service Elementary Education Majors' Attitudes toward Mathematics: A Semantic Differential

Carmen M. Latterell and Janelle L. Wilson

INTRODUCTION

Ability in mathematics is viewed by nearly everyone as an essential life skill (Boaler, 2008; Gallup, 2005; Peker & Mirasyedioğlu, 2008; Vásquez-Colina, Gonzalez-DeHass, & Furnter, 2014; Zollman, 2012). President Obama's campaign Educate to Innovate was intended to help students in science, technology, engineering, and mathematics (STEM), and in every one of President Obama's State of the Union Addresses, he spoke on the need for mathematical skills. In his final State of the Union Address, he stated that every student needs "computer science and math classes that make them job-ready on day one" (Obama, 2016).

Yet many people are not successful at mathematics and do not like mathematics (Furner & Gonzalez-DeHass, 2011; Peker & Mirasyedioğlu, 2008; Zollman, 2012). Many researchers believe that this dislike of mathematics begins in elementary school and is attributable, in part, to a dislike of mathematics prevalent among many elementary school teachers. It is believed, then, that knowingly or unknowingly, teachers pass along unfavorable attitudes and feelings about mathematics to their young students thus precipitating a cycle of negative views of mathematics (Brady & Bowd, 2005; Harper & Daane, 1998; Jackson & Leffingwell, 1999).

A positive attitude toward mathematics learning is needed in order to be successful at it (Aschcraft, 2002; Cornell, 1999; Rameau & Louime, 2007; Geist, 2010; Kalder & Lesik, 2011; Ma, 1999; Popham, 2008; Relich, Way, & Martin, 1994; Vinson, 2001). Elementary teachers influence the forming of the earliest attitudes of their students (Bekdemir, 2010; Gresham, 2007; Relich, Way, & Martin, 1994; White, Way, Perry, & Southwell, 2005; Ma, 1999). Of course, elementary teachers themselves form attitudes before becoming elementary teachers (Bekdemir, 2010; Bobis & Cusworth, 1994; Davies & Savell, 2000; Grootenboer, 2002; McAnallen, 2010; Uusimaki & Nason, 2004). These attitudes influence the way that they teach (Ball, 1990; Haciomeroglu, 2013; Ramey-Gossert & Schroyer, 1992; Swars et al., 2007; Thompson, 1992; Wilson & Cooney, 2002; Ma, 1999).

Research has shown that pre-service elementary teachers' attitudes toward mathematics are at best not positive and at worse quite negative (Davies & Savell, 2000; Grootenboer, 2002; Relich, Way, & Martin, 1994; Tsao, 2014). In addition, many pre-service elementary teachers have anxiety about mathematics (Bekdemir, 2010; Bursal & Paznokas, 2006; Gresham, 2007; Harper & Daane, 1998; McAnallen, 2010; Tsao, 2014). This anxiety is part of a cycle that is set in motion during elementary school as the anxiety is often created by pre-service elementary teachers' own early mathematics experiences (Brady & Bowd, 2005; Hadfield & Lillibridge, 1991; Harper & Daane, 1998; Vinson, 2001).

Pre-service elementary teachers give a variety of reasons for their negative attitudes, including that they had poor teachers (Bekdemir, 2010; Cornell, 1999; Uusimaki & Nason, 2004) and they could not keep up with the class (Cornell, 1999). In addition, a significant number of elementary students do not believe that mathematics is useful (Kloosterman & Cougan, 1994).

WHAT CAN BE DONE TO BREAK THE NEGATIVE CYCLE?

Two main themes emerge on how to break this cycle. The first is to increase pre-service elementary teachers' awareness of their feelings about and attitudes toward mathematics (in particular, any math anxiety they have experienced) and how those feelings and attitudes may affect the way they teach the subject. Elementary education teacher programs could take the lead in incorporating anxiety awareness into their curriculum. Awareness of the problem may also lead to numerous strategies that the pre-service teacher might employ to overcome the problem, including anxiety-reducing techniques (such as visualizing success and deep breathing) and better preparation skills.

The other theme that has emerged in the research is that mathematics curriculum and pedagogy should change. The National Council of Teachers of Mathematics outlines curriculum and pedagogy in their standards (National Council of Teachers of Mathematics, 1989, 1991, 1995, 2000) with the intention of reducing anxiety and increasing positive attitudes.

This reform-oriented approach *decreases the concentration on procedures* and *increases the concentration on concepts*. More discussion and real-world applications are incorporated, along with experiential projects. In addition, students do mathematical inquiry, a discovery approach to mathematics. The aim in reform mathematics is for students to understand the "why" behind their work and, at times, to focus less on computational mathematics.

Just because the National Council of Teachers of Mathematics recommends a certain curriculum and approach, however, does not guarantee that teachers teach in that manner. In fact, most research suggests that reform is not being implemented (Ross, McDougall, & Hogaboam-Gray, 2002). On the other hand, some research suggests that teachers want to teach in reform manners but do not know how (Frykholm, 1996).

In a 1999 study, the researchers found "evidence of 'reform-oriented' mathematics practice in each of the 25 classrooms [they] studied" (Spillane & Zeuli, 1999, p. 18) but their patterns of practice did not support consistent and thorough reform. In a similar manner, while piloting an instrument for measuring reform, researchers found teachers with a high level of reform (Ross, McDougall, Hogaboam-Gray, & LeSage, 2003).

It is beyond the scope of this research to determine the current level of implemented reform. However, it seems reasonable to claim that there is both interest in mathematics education reform and a mathematics education culture that supports the idea of reform and that this culture has been increasing since around 1990. Although the current project will not claim a connection between reform and the lessening of mathematics anxiety, it behooves the research community to continue to assess whether the attitudes of elementary teachers are changing. Is it possible that we are beginning to break the cycle of negativity?

RESEARCH QUESTIONS

1. On a scale from very negative to very positive, what is the rating of pre-service elementary education majors in regard to mathematics?
2. Do pre-service elementary education majors' semantic differential scores match their written narratives regarding attitudes and experiences with mathematics?

METHOD

There are two goals, then, of this study: (1) create a numeric rating of pre-service elementary education majors' attitudes toward mathematics, and (2) compare this numeric rating to their narratives. The key components of the study include the sample, the instrument (which was created for this study), and the analysis. The sample is described first. But before these discussions, a brief note about the Institutional Review Board (IRB) process at both authors' university: The IRB at the university reviews all research projects involving human participants to ensure that all participants are protected and that informed uncoerced consent occurs. The authors of this study applied for and received IRB approval.

SAMPLE

As this study's focus is on the attitudes of pre-service elementary education majors, the authors collected a purposive sample for the study. The sample consisted of 54 students enrolled in a mathematics content course in the spring of 2016 at a mid-size university in the Midwest. The course, Mathematics for Elementary Education Majors II, is the second in a two-part series that only elementary education majors take. The content of this particular course was geometry, probability, and statistics. One of the authors was the instructor of the course. The students were given informed consent forms for the use of their completed instrument.

Filling out the instrument was a class assignment. Therefore, all students completed the instrument. However, no extra credit for offering permission to use their filled-out instrument for this study was given nor was any penalty applied for not offering such permission. As it turns out, all students granted permission.

INSTRUMENT

The instrument had two parts: a semantic differential and a fill-in-the-blank narrative. The semantic differential is a rating scale that the authors developed. The section below describes this development in order to offer content validity in the use of the instrument. The fill-in-the-blank narrative was also self-developed and was intended to allow the subjects to expand on their thoughts. This is also described in more detail below.

Semantic Differential

A semantic differential is a rating scale made up of polar adjectives to represent the connotative meaning of a concept. The rating scale was developed by Charles E. Osgood in the 1950s and used extensively during the 1960s (Osgood, Suci, & Tannenbaum, 1957). It is still used today in disciplines such as psychology and sociology (Ploder & Eder, 2015) and is "especially suitable for measuring emotional and behavioural aspects of the attitude" (Chráska & Chrásková, 2016, p. 821).

This emotional aspect of pre-service elementary teachers' attitudes toward mathematics is precisely what the authors wanted to measure. In particular, the desire was to understand if such a rating from very negative to very positive actually did match pre-service elementary education majors' feelings about, experiences with, and attitudes toward mathematics.

The authors chose to use this measurement tool because the primary objective was to ascertain the subjects' attitude toward mathematics. To develop this scale, students enrolled in Mathematics for Elementary Education Majors II ($N = 54$) were asked to list five adjectives that describe the concept "math." The adjectives listed by over twenty of the students each were *challenging* ($n = 24$) and *interesting* ($n = 22$). Five adjectives were listed by over ten students each: *confusing* ($n = 17$), *important* ($n = 15$), *time consuming* ($n = 15$), *fun* ($n = 12$), and *rewarding* ($n = 11$). *Complicated* and *difficult* were each listed nine times, with *hard* ($n = 8$) and *complex* ($n = 7$) having a bit less frequency. *Exciting* and *frustrating* were each listed five times, and *essential*, *helpful*, and *stressful* were each listed four times.

Although several adjectives were listed two or three times, most of the remaining adjectives were only listed once (over 40 adjectives in total). The most popular adjectives were used to form the semantic differential (see Table 3.1).

It is conceivable that other researchers could arrive at a semantic differential in the same way—i.e., inductively. Indeed, it would be interesting to see how similar (or different) such semantic differentials would be. Alternatively, other researchers could simply adopt the semantic differential that was developed and used in this study and in this way replicate the study.

Narrative

The narrative was a fill-in-the-blank questionnaire that can be found in the appendix. It asked participants to report their feelings and had participants recall particular events that led to those feelings. It ended with a fill-in-the-blank section that asked for overall feelings and abilities regarding mathematics. The purpose of the narrative was to compare the results of each student to that student's semantic differential rating. If there was a strong match, then the validity of the semantic differential was enhanced—or possibly the semantic differential was all that was needed. Other researchers who wish to replicate this study could readily adopt this narrative.

Table 3.1.   Rate "Math" by placing an X in one of the boxes for each dimension.

Challenging	__: __: __: __: __ : __ : __	Cinch
Boring	__: __: __: __: __ : __ : __	Interesting
Confusing	__: __: __: __: __ : __ : __	Clear
Unimportant	__: __: __: __: __ : __ : __	Important
Time consuming	__: __: __: __: __ : __ : __	Quick
Dreadful	__: __: __: __: __ : __ : __	Fun
Unrewarding	__: __: __: __: __ : __ : __	Rewarding
Complicated	__: __: __: __: __ : __ : __	Uncomplicated
Difficult	__: __: __: __: __ : __ : __	Easy
Complex	__: __: __: __: __ : __ : __	Simple
Unstimulating	__: __: __: __: __ : __ : __	Exciting
Frustrating	__: __: __: __: __ : __ : __	Encouraging
Not necessary	__: __: __: __: __ : __ : __	Essential
Stressful	__: __: __: __: __ : __ : __	Pleasant

Analysis

To score the semantic differential, each blank was counted using a scale of 1 through 7, with 1 at the negative end. An average was found for each individual and recorded on the following scale: [0, 2) very negative, [2, 3) negative, [3, 4) moderately negative, [4, 5) neutral, [5, 6) moderately positive, [6, 7) positive, 7 very positive.

To analyze the narratives, the two researchers separately read each narrative to find themes and key words. The researchers followed a grounded theory approach in coding the data, allowing themes to emerge (Charmaz, 2006). The researchers then met to discuss the level of agreement on themes. Although each researcher used different wording, there was 100% agreement on the themes. Once a wording for the themes was agreed on, the researchers again separately tallied which narratives met each theme. This allowed the researchers to calculate an interrater reliability score to measure the level of agreement. The original interrater reliability statistic was 96% and reached 100% after discussion.

RESULTS

The results are given in two parts: the semantic differential results and the narrative results. For the semantic differential results, the class averages are given along with a method of providing a summary of individual averages. For the narrative results, a summary of the responses is provided along with particular individual responses.

Semantic Differential Results

As stated earlier, the sample consisted of an entire class ($N = 54$) filling out the narrative and taking the semantic differential. The class averages on the semantic differential are given in Table 3.2. Seven of the items averaged in the neutral category. Three of the items averaged in the moderately negative category. These three concerned the amount of time that math takes as well as how complicated or complex it is.

Actually, even students who like mathematics a great deal might rate it as time consuming or even as complex. Two of the items averaged in the moderately positive category. These items concerned how interesting and rewarding mathematics is. Finally, two items averaged in the positive category, and these were in regard to the essential nature and importance of mathematics.

Table 3.2. Class Averages on Semantic Differentials

Adjective Pairs	Mean*	Standard Deviation
Challenging versus *Cinch*	3.759	1.464
Boring versus *Interesting*	4.648	1.231
Confusing versus *Clear*	4.13	1.555
Unimportant versus *Important*	6.019	1.073
Time Consuming versus *Quick*	3.37	1.496
Dreadful versus *Fun*	4.426	1.075
Unrewarding versus *Rewarding*	5.389	1.14
Complicated versus *Uncomplicated*	3.389	1.435
Difficult versus *Easy*	3.833	1.424
Complex versus *Simple*	3.296	1.449
Unstimulating versus *Exciting*	4.315	1.256
Frustrating versus *Encouraging*	3.907	1.321
Not Necessary versus *Essential*	5.944	1.089
Stressful versus *Pleasant*	3.759	1.400

*The furthest left was rated 1 and the furthest right was rated 7. Thus 1–3 could be considered negative, 4 neutral, and 5–7 positive.

Each individual also had an average, and these were categorized between very negative and very positive. The total percentage of students in each category is given in Table 3.3. Only 5% ($n = 3$) of the students were categorized in the negative category. The remaining 95% ($n = 51$) of the students were categorized as moderately negative (30%, $n = 16$), neutral (41%, $n = 22$), or moderately positive (24%, $n = 13$). Thus none of the students were categorized as very negative, positive, or very positive. The average of all students was 4.29, which is in the neutral range.

After making these calculations, the researchers were concerned that some items considered negative might not be perceived as negative by participants. The semantic differential was given to a class of math majors, and math majors often view mathematics as challenging, time consuming, complicated, difficult, and complex. However, if the math majors were told to rate "elementary" math, they might give different ratings. The decision was made to exclude all five of these categories and calculate an average based on the remaining nine categories (the unequivocally negative terms: *boring, confusing, unimportant, dreadful, unrewarding, unstimulating, frustrating, not necessary,* and *stressful*).

The average of all students using only the nine categories was 4.73. According to these results, no students categorized as negative, and 6% ($n = 3$) of the students categorized as positive. The remaining 94% ($n = 51$) were moderately negative (13%, $n = 7$), neutral (48%, $n = 26$), or moderately positive (33%, $n = 18$).

Table 3.3. Individual Averages

Scale	Percentage of Students	Percentage of Students When 5 Items Were Not Included
Very Negative	0	0%
Negative	5%	0%
Moderately Negative	30%	13%
Neutral	41%	48%
Moderately Positive	24%	33%
Positive	0	6%
Very Positive	0	0%

Narrative Results

In addition to the semantic differential, participants filled in the blanks in a narrative. Table 3.4 gives a summary of responses as well as sample words used. In the narrative, participants were asked for their earliest memory of doing mathematics and how they felt about mathematics then. Second, they were asked for another memory of doing mathematics that occurred at a later time and their feeling about mathematics at that time. Finally, they were asked for their current feelings on doing mathematics and how good at mathematics they considered themselves to be.

Earliest Memory

Participants were first asked for their earliest memory of doing mathematics. The majority of the responses represented positive memories (70%, $n = 38$).

Fifteen participants used words that describe negative feelings and one participant used the word *apathetic*. Participants were then asked the degree to which they cared for mathematics at the time of this earliest memory. Seventy percent ($n = 38$) again gave positive words such as *liked* and *loved*. Another twelve participants gave negative words such as *disliked*. The researchers separated another four responses in which the participants were neutral.

Later Memory

Participants were then asked to think of an event later than their earliest memory and describe how it changed or did not change their feelings about mathematics. Only about 31% ($n = 17$) of the participants reported continued positive feelings. Out of these, some reported feeling that they always did well with mathematics, while quite a few reported that they were usually better than other people were. This idea of being better than others was an important motivation for some of the participants. Other positive comments included having helpful teachers and teachers who made things fun.

About 9% ($n = 5$) of the participants reported negative feelings about mathematics such as feeling that others were always better than them or getting behind and not catching up. Others reported that they had a bad teacher.

The majority of participants (the remaining 60%, $n = 32$) reported that mathematics sometimes went well and sometimes did not. This was the first indication that mathematics has an up-and-down nature to it.

Current State

Participants were then asked what factors seemed to account for their changing view of mathematics. Over 75% of the participants said that the number one factor was teachers. Five participants talked about their negative feelings about particular curriculum material.

Finally, the participants were asked to offer concluding comments on how well they currently liked mathematics and how good at it they thought they were. About 54% ($n = 29$) gave a positive view of how well they liked mathematics. Note that this is down from the 70% that began liking mathematics. However, only 11% ($n = 6$) gave a negative word for liking mathematics, with 17% ($n = 9$) feeling neutral about it and another 18% ($n = 10$) sometimes liking mathematics and sometimes not.

In terms of how good at mathematics the participants were, about 55% ($n = 30$) felt that they were good at mathematics and 4% ($n = 2$) felt they were bad at mathematics. The remaining 41% ($n = 22$) felt they were average.

Table 3.4. Summary of Responses with Sample Words

	Positive	Negative	Neutral	Positive at times and Negative at Times
Earliest memory of doing mathematics	n = 38 Good, Confident, Happy, Excited, Curious, Content, Interested	n = 15 Confused, Annoyed, Unclear, Incompetent, Nervous, Anxious, On edge	n = 1 Apathetic	
Degree to which participant cared for mathematics	n = 38 Liked, Loved, Enjoyed, Really liked	n = 12 Disliked, Didn't care for it	n = 4 Neutral, Indifferent	
Later event	n = 17 They were better than others, It always went well, Good teacher	n = 5 Others were always better, Got behind and could not catch up, Bad teacher		n = 32 Teacher helped when fell behind, Subjects changed, Got harder
How well they currently like mathematics	n = 29 Really like, Enjoy, Am pretty okay with math, Love doing math	n = 6 Frustrated, Not a big fan, Nervous	n = 9 Indifferent, Neutral, Unbiased, Neither like nor dislike	n = 10 Enjoy parts of it, Enjoy algebra, Dislike the others
How good at mathematics they currently report	n = 30 Very good, Good, Pretty good	n = 2 Poor	n = 22 Average	

DISCUSSION

Our first research question asked what the rating of pre-service elementary education majors is in regard to mathematics. The averages from the semantic differential represent pre-service elementary education majors who were mostly neutral or somewhat positive. Only 30% ($n = 16$) of the averages were moderately negative, and only 5% ($n = 3$) were negative. This seems to indicate that the pre-service elementary education teachers in this sample were not as negative as previous research states (Davies & Savell, 2000; Grootenboer, 2002: Relich, Way, & Martin, 1994; Tsao, 2014).

In addition, on questions of importance and necessity, the pre-service elementary teachers rated mathematics positively. This too is a result that is not consistent with what previous research has shown (Kloosterman & Cougan, 1994). Finally, if certain items are not counted (such as the time-consuming nature of mathematics) because they may or may not be seen as negative, then the pre-service elementary education majors rated higher with 13% ($n = 7$) moderately negative and everyone else rating higher.

Our second research question asked whether the semantic differential scores matched the written narratives that pre-service elementary teachers gave. The narrative revealed several themes. A consistent theme that ran through a majority of responses was that mathematics sometimes goes well and sometimes does not. For example, one respondent noted falling behind in mathematics then having a good teacher and doing better but then not doing as well in high school.

A couple of respondents noted that mathematics got harder and was then less enjoyable. Another respondent wrote about understanding mathematics early on but experiencing it as more difficult in high school, although still doing well. A couple of respondents indicated that they did better in some mathematics subjects than others (e.g., "loved algebra, hated calculus"). One respondent reported having a great teacher in third grade but a boring teacher in fourth grade. One respondent reported struggling in mathematics until 10th grade. Two respondents said that as mathematics got harder, it became less enjoyable.

When respondents wrote about negative experiences with teachers, they wrote both about not understanding mathematics well primarily due to poor teachers and about feeling shamed by their teachers. One respondent indicated feeling too nervous to ask for help.

Most of these themes would not have been known without the narrative. The theme of the up-and-down nature of mathematics was not obvious in the semantic differential. However, the theme that mathematics is useful and important was not obvious in the narrative. The narrative revealed the same neutral to moderately positive view that the semantic differential did. Further, the narrative demonstrated that the students were more positive earlier on. Of course, now that certain themes stood out, they could be represented on the semantic differential (e.g., "always the same" on one pole and "up and down" on the other).

In short, the semantic differential revealed the general state of the pre-service elementary teachers and was quick to both administer and analyze while the fill-in-the-blank narrative part of the study instrument allowed participants to provide details that better contextualized their mathematics trajectories. Taken together, this two-part instrument gave us a more complete view.

CONCLUSIONS

Due to the ease, quickness, and accuracy of the semantic differential, we recommend its use in teacher education programs to continue to determine the attitudes of pre-service elementary teachers. It is important to break the cycle of negativity. The results of our study suggest hopeful signs that the negativity is indeed lessening. We continue to see the importance of teachers in influencing attitudes, but also significant is the nature of mathematics to be both up and down. We need to work with students at all levels to understand that this is a part of learning mathematics and to encourage students to continue on the path even when it is "down" because the "up" part of mathematics is sure to come as well.

APPENDIX

Questionnaire Instructions: Please fill in the blanks below.

In my earliest memory of doing mathematics, I remember feeling _____ (e.g., happy, unhappy, nervous, etc.) and at that point I _____ (e.g., liked, loved, disliked, hated, etc.) mathematics.

As time went on, I remember this experience with math: _____

_____

_____

_____

_____

_____

And it _____ (e.g., confirmed, changed, etc.) my view of math.

I think that _____ (e.g., teachers, family, course material or topic, friends, etc.) has affected my view of mathematics in a _____(e.g., positive, negative, etc.) way.

Overall I _____
_____ (enter how you feel about) math.

In fact, I identify as _____ (enter your level of ability, e.g., superior, very good, good, average, poor) in math.

REFERENCES

Ashcraft, M. H. (2002). Math anxiety: Personal, education, and cognitive consequences. *Current Directions in Psychological Science, 11*(5), 181–185.

Ball, D. L. (1990). The mathematical understanding that prospective teachers bring to teacher education. *Elementary School Journal, 90*, 449–465.

Bekdemir, M. (2010). The pre-service teachers' mathematics anxiety related to depth of negative experiences in mathematics classroom while they were students. *Educational Studies in Mathematics, 75*, 311–328.

Boaler, J. (2008). *What's math got to do with it? Helping children learn to love their least favorite subject—And why it's important for America.* New York, NY: Penguin.

Bobis, J., & Cusworth, R. (1994). Teacher education: A watershed for preservice teachers' attitudes toward mathematics. In G. Bell, B. Wright, N. Leeson, & J. Geake (Eds.), *Challenges in mathematics education: Constraints on construction.* Proceedings of the 17th Annual Conference of the Mathematics Education Research Group of Australasia, vol. 1, pp. 113–120. Lismore, New South Wales, Australia: MERGA.

Brady, P., & Bowd, A. (2005). Mathematics anxiety, prior experience and confidence to teach mathematics among pre-service education students. *Teachers and Teaching, 11*(1), 37–46.

Bursal, M., & Paznokas, L. (2006). Mathematics anxiety and pre-service elementary teachers' confidence to teach mathematics and science. *School Science and Mathematics, 106*(4), 173–179.

Charmaz, K. (2006). *Constructing grounded theory: A practical guide through qualitative analysis.* Thousand Oaks, CA: Sage Publications.

Chráska, M., & Chrásková, M. (2016). Sematic differential and its risks in the measurement of students' attitudes. *Procedia—Social and Behavioral Sciences, 217*, 820–829.

Cornell, C. (1999). I hate math! I couldn't learn it, and I can't teach it! *Childhood Education, 75*(4), 225–230.

Davies, N., & Savell, J. (2000). "Maths is like a bag of tomatoes": Student attitudes upon entry to an early years teaching degree. Paper presented at the Teacher Education Forum of Aotearoa New Zealand Conference, Christchurch, New Zealand.

Educate to Innovate. (2010). *Educate to innovate: Overview.* Washington, DC: Author. www.whitehouse.gov/issues/education/educate-innovate

Frykholm, J. A. (1996). Pre-service teachers in mathematics: Struggling with the standards. *Teaching and Teacher Education, 12,* 665–681.

Furner, J. M., & Gonzalez-DeHass, A. (2011). How do students' mastery and performance goals relate to math anxiety? *Eurasia Journal of Mathematics, Science & Technology Education, 7*(4), 227–242.

Gallup, G. H. (2005). Math problematic for U. S. teens: Youth survey pool. www .gallup.com/poll/16360/math-problematic-us-teens.aspx

Geist, E. (2010). The anti-anxiety curriculum: Combating math anxiety in the classroom. *Journal of Instructional Psychology, 37*(1), 24–31.

Gresham, G. (2007). A study of mathematics anxiety in pre-service teachers. *Early Childhood Education Journal, 35*(2), 181–188.

Grootenboer, P. J. (2002). Affective development in mathematics: A case of two preservice primary school teachers. In B. Barton, K. Irwin, M. Pfannkuch, & M. Thomas (Eds.), *Mathematics education in the South Pacific.* Proceedings of the 25th Annual Conference of the Mathematics Education Research Group of Australasia (vol. 1, pp. 318–325). Sydney, New South Wales, Australia: MERGA.

Haciomeroglu, G. (2013). Mathematics anxiety and mathematical beliefs: What is the relationship in elementary pre-service teachers? *Issues in the Undergraduate Mathematics Preparation of School Teachers, 5,* 1–9.

Hadfield, O. D., & Lillibridge, F. (1991). *A hands-on approach to the improvement of rural elementary teacher confidence in science and mathematics.* Nashville, TN: Annual National Rural Small Schools Conference. ERIC Document Reproduction Service no. ED 334082.

Harper, N. W., & Daane, C. J. (1998). Causes and reduction of math anxiety in pre-service elementary teachers. *Action in Teacher Education, 19*(4), 29–38.

Jackson, C. D., & Leffingwell, R. J. (1999). The role of instructors in creating math anxiety in students from kindergarten through college. *Mathematics Teacher, 92*(7), 583–586.

Kalder, R. S., & Lesik, S. A. (2011). A classification of attitudes and beliefs toward mathematics for secondary mathematics pre-service teachers and elementary pre-service teachers: An exploratory study using latent class analysis. *Issues in the Undergraduate Mathematics Preparation of School Teachers, 5.* www.k-12prep .math.ttu.edu/journal/journal.shtml

Kloosterman, P., & Cougan, M. C. (1994). Students' beliefs about learning school mathematics. *Elementary School Journal, 94*(4), 375–388.

Ma, X. (1999). A meta-analysis of the relationship between anxiety toward mathematics and achievement in mathematics. *Journal for Research in Mathematics Education, 30*(5), 520–540.

McAnallen, R. R. (2010). Examining mathematics anxiety in elementary classroom teachers (Doctoral Dissertation). University of Connecticut.

National Council of Teachers of Mathematics. (1989). *Curriculum and evaluation standards for school mathematics*. Reston, VA: Author.

National Council of Teachers of Mathematics. (1991). *Professional standards for teaching mathematics*. Reston, VA: Author.

National Council of Teachers of Mathematics. (1995). *Assessment standards for school mathematics.* Reston, VA: Author.

National Council of Teachers of Mathematics. (2000). *Principles and standards for school mathematics.* Reston, VA: Author.

Obama, B. (2016). State of the Union Address. Retrieved from www.whitehouse.gov /the-press-office/2016/01/12/remarks-president-barack-obama-prepared-delivery -state-union-address

Osgood, C. E., Suci, G., & Tannenbaum, P. (1957). *The measurement of meaning.* Urbana, IL: University of Illinois Press.

Peker, M., & Mirasyedioğlu, S. (2008). Pre-service elementary school teachers' learning styles and attitudes towards mathematics. *Eurasia Journal of Mathematics, Science & Technology Education, 4*(1), 21–26.

Ploder, A., & Eder, A. (2015). Semantic differential. *International Encyclopedia of the Social & Behavioral Sciences, 21*, 563–571.

Popham, W. J. (2008). Timed tests for tykes? *Educational Leadership, 65*(8), 86–87.

Rameau, P., & Louime, C. (2007). Mathematics phobia: Are the mathematical sciences a pothole in the road of life? *Current Science, 93*(11), 1481–1482.

Ramey-Gassert, L., & Shroyer, M. G. (1992). Enhancing science teaching self-efficacy in preservice elementary teachers. *Journal of Elementary Science Education, 4*, 26–34.

Relich, J., Way, J., & Martin, A. (1994). Attitudes to teaching mathematics: Further development of a measurement instrument. *Mathematics Education Research Journal, 6*(1), 56–69.

Ross, J. A., McDougall, D., & Hogaboam-Gray, A. (2002). Research on reform in mathematics education, 1993–2000. *Alberta Journal of Educational Research, 48*(2), 122–138.

Ross, J. A., McDougall, D., Hogaboam-Gray, A., & LeSage, A. (2003). A survey measuring elementary teachers' implementation of standards-based mathematics. *Journal for Research in Mathematics Education, 34*(4), 344–363.

Spillane, J. P., & Zeuli, J. S. (1999). Reform and teaching: Exploring patterns of practice in the context of national and state mathematics reforms. *Educational Evaluation and Policy Analysis, 21*, 1–27.

Swars, S., Hart, L. C., Smith, S. Z., Smith, M. E., & Tolar, T. (2007). A longitudinal study of elementary pre-service teachers' mathematical beliefs and content knowledge. *School Science and Mathematics, 107*(9), 325–335.

Thompson, A. (1992). Teacher's beliefs and conceptions: A synthesis of the research. In D. A. Grouws (Ed.), *Handbook of research on mathematics teaching and learning* (pp. 127–146). New York, NY: Macmillan.

Tsao, Y. L. (2014). Attitudes and beliefs toward mathematics for elementary pre-service teachers. *US–China Education Review B, 4*(9), 616–626.

Uusimaki, L., & Nason, R. (2004). Causes underlying pre-service teachers' negative beliefs and anxieties about mathematics. *Proceedings of the 28th Conference of the International Group for the Psychology of Mathematics Education, 4*, 369–376.

Vásquez-Colina, M. D., Gonzalez-DeHass, A. R., & Furnter, J. M. (2014). Achievement goals, motivation to learn, and mathematics anxiety among pre-service teachers. *Journal of Research in Education, 24*(1), 38–52.

Vinson, B. M. (2001). A comparison of preservice teachers' mathematics anxiety before and after a methods class emphasizing manipulatives. *Early Childhood Education Journal, 29*(2), 89–94.

White, A. L., Way, J., Perry, B., & Southwell, B. (2005). Mathematical attitudes, beliefs and achievement in primary pre-service mathematics teacher education. *Mathematics Teacher Education and Development, 7*, 33–52.

Wilson, M., & Cooney, T. J. (2002). Mathematics teacher change and development. In G. C. Leder, E. Pehkonen, & G. Torner (Eds.), *Beliefs: A hidden variable in mathematics education?* (pp. 127–147). Dordrecht, The Netherlands: Kluwer.

Zollman, A. (2012). Learning for STEM literacy: STEM literacy for learning. *School Science and Mathematics, 112*(1), 12–19.

Chapter 4

Addressing Math Phobia at Its Source

A Case Study

Melinda (Mindy) Eichhorn and Courtney Lacson

INTRODUCTION

How can we transform our approach to teaching early and elementary mathematics to address the national problem of mathematics phobia? McCray (2016) has argued, "To stop creating a preponderance of students who are certain they are not and never will be 'good at math,' we must fundamentally transform how math is taught in the early years" (par. 5).

Strong academic standards and quality curriculum are important to mathematics instruction, yet skilled teachers, who are confident in their mathematical abilities, must engage and support students to learn mathematics. Early childhood and elementary education teachers need mathematical knowledge for teaching, or mathematical reasoning, insight, understanding, and skill. They must also be fluent and have a deep understanding of mathematical vocabulary and terms (Ball, Hill, & Bass, 2005).

Teachers of mathematics analyze students' responses and errors. Additionally, teachers create a safe space for exploring math ideas and taking risks as they act as a mathematical facilitator rather than an authoritarian who insists every student complete the problem the way they (the teacher) solved it. Teachers develop five strands of math proficiency in their students: conceptual understanding, procedural fluency, strategic competence, adaptive reasoning, and productive disposition (National Research Council, 2001).

Yet pre-service teachers typically enter math methods and content courses with a limited and traditional view of math—one that focuses on computation and following an established procedure to obtain an answer (Koestler, 2012). Early elementary teachers typically teach math the way they were taught math (McCray, 2016; Remillard, 2000). They were taught to follow the procedure and obtain the one correct answer (e.g., "Don't ask why, just invert and multiply"). As math becomes more abstract, it can appear to be full of magical tricks without making much sense. Some worked hard to learn rules and memorize steps yet never gained confidence or automaticity in mathematics.

Perhaps math became confusing and scary and students developed "math helplessness" (McCray, 2016, par. 3). As pre-service teachers enter math content and methods courses, they might be hesitant and anxious about teaching others math because they do not feel confident in mathematical conceptual understanding (Koestler, 2012).

How do previous experiences in math affect our beliefs about math or lead to "math trauma" (Boaler, 2016, p. xii)? How do our views of math influence our teaching of math? How can new teachers instill a growth mindset approach in their elementary students in regard to math if they have not experienced a growth mindset in math themselves? In this study, we examine the beliefs, perceptions, and experiences of pre-service teachers and recent alumni at a private, liberal arts college in the Northeast to determine if early childhood and elementary teacher candidates and alumni have beliefs about math that are aligned with the literature such as low confidence in math and math anxiety. We outline changes to the current math program in early childhood and elementary education based on these findings.

FEMALE TEACHERS AND MATH

More than 90% of early elementary teachers in the United States are female (Beilock, Gunderson, Ramirez, & Levine, 2010). Early childhood and elementary pre-service and current teachers may feel anxious about math due to their own math trauma experiences in school and through cultural messages about their potential in math (Boaler, 2016). Female early childhood and elementary majors may avoid high-level math courses or perform poorly on math assessments because female elementary teachers, and girls in general, are perceived as having weak math skills. Females may feel that they confirm this negative stereotype. This is called a *stereotype threat.*

Stereotype threats, or implications of a social identity, can also affect intellectual functioning in math and impair performance (Steele, 2010). Elementary teachers may in turn believe that they are not a *math person.* Math-anxious individuals can impact their students' math achievement—more so for female students that for male students (Beilock et al., 2010; Boaler, 2016).

MATHEMATICAL MINDSETS

According to Dweck (2006), we all have a core belief, or a *mindset*, about how we learn. We can have a fixed mindset, believing that we can learn things but not change our intelligence, or a growth mindset, believing that with hard work and perseverance we can learn and achieve at high levels (Boaler, 2016). Boaler asserted that you can have a growth mindset in everything except math due to strong and negative ideas about math. Therefore, special attention must be given to cultivating a mathematical mindset. When teachers adopt a growth mindset approach in mathematics by emphasizing strategies and persistence, student motivation and self-expectation are influenced and stereotype threats are reduced (CAST, 2016; Busch, 2017).

Many students have experienced math trauma due to poor math teaching and the impact of pervasive cultural messages that math is hard and only some people can do it well (Boaler, 2016). Math is conveyed as a difficult subject and in a negative light through subtle messages in the media (e.g., "Math is for nerds"). However, with the right teaching and support, all children can achieve in math (Boaler, 2016).

Speed has been wrongfully regarded as a characteristic of being good or smart at math. Instead, high-achieving students in math think conceptually and are flexible with numbers rather than rely on standard methods or procedures (Boaler, 2015; Gray & Tall, 1994). To think deeply and carefully about math does not necessarily mean that you think the fastest. The focus on speed in math can be an attributing factor to students' aversion to math. By giving timed tests of math facts in isolation, teachers may be creating math anxiety and adding to the belief that you have to be fast in order to be good at math (Boaler, 2016). Math anxiety is not poor math ability but the emotional response that gets in the way of showing true math ability (Beilock et al., 2010).

Mathematicians in the real world think deeply and carefully about math, and this does not often happen quickly. Depth of thinking, not speed, is more reflective of math in the real world. Students may believe that if they do not understand the math and obtain the answer in a few seconds, they will never understand it. Successful mathematicians, on the other hand, have the ability to concentrate deeply for extended periods and consider several solution pathways. Mathematicians also pose interesting questions (Boaler, 2016). While this is important in mathematical work, it is absent from many math classrooms.

TEACHER IDENTITIES

The way teachers learn and teach math is shaped by their sense of self, knowledge, beliefs, dispositions, interests, and orientation toward math (Spillane, 2000). Teachers' beliefs about math are strongly influenced by immediate and extended family, teaching traditions, and atypical teaching episodes (Beijaard, Meijer, & Verloop, 2004). Listening to teachers' stories about their mathematical identity helps us to understand their beliefs, motivations, and perspectives (Drake, Spillane, & Hufferd-Ackles, 2001).

Pre-service teachers and current teachers need to engage in self-examination to understand and challenge their own beliefs, attitudes, and actions toward math as well as the way these beliefs will affect their teaching of math and their impressions of their students (Koestler, 2012; Remillard, 2000; Turner, 2013). How have their previous math experiences provided engaging and challenging activities in math? Have their previous teachers praised them for their persistence and perseverance? Overall, what factors have affected their experiences in mathematics?

This study examines pre-service teachers' perception of math as well as the parts of their coursework in early childhood and elementary education that have been most helpful as they engage in mathematical work with students. We embarked on this study to find out how early childhood and elementary majors think about math, how their previous experiences have shaped their view of math, and how the trends among students and alumni can be generalized to trends across the United States. We also used this research to adjust coursework and assignments for pre-service teachers.

RESEARCH QUESTIONS

The study focused on the following research questions:

1. How do prior experiences in mathematics shape teacher candidates' and current teachers' beliefs about and perceptions of mathematics and mathematics teaching?
2. Which parts of their coursework in their teacher-prep program and classroom experiences have been most helpful as they engage in mathematical work with early childhood and elementary students?

THEORETICAL FRAMEWORK

The theoretical framework for our study is social constructionism, which is focused on the way people "describe, explain, or otherwise account for the world (including themselves) in which they live" (Gergen, 1985, p. 266). Social constructionism encourages us to question our assumptions and perceptions (Burr, 1995). Because we are learners, there is variability in our dispositions and beliefs about math, which are rooted in our identities. By listening to students' and teachers' stories about their mathematical identity we can begin to understand their beliefs, motivations, and perspectives (Drake et al., 2001). Belief structures are a product of an individual's experiences with mathematics and are slow to form and difficult to change.

We examined the data in the study through a social constructionism framework with an overlapping lens of fixed and growth mindsets specific to math. We studied how current students and graduates describe and interpret their experiences in math and what they have done in response to these early experiences. It is helpful to evaluate pre-service teachers' math identities in math methods and math content classes to acknowledge the variability in beliefs about math and begin to help cultivate a growth mindset toward math.

METHODS

This case study followed a mixed-methods design of a teacher education program. The researchers examined the math attitudes and beliefs of the students and recent alumni.

Setting

This mixed-methods study was based at a small, private, liberal arts college in the northeastern United States. This study was part of a summer research collaboration between a full-time faculty member and an undergraduate elementary education major.

Participants and Procedures

Due to the limited time period available for data collection and analysis, we employed a convergent mixed-methods approach, collecting both qualitative and quantitative data concurrently. We then integrated the information in our interpretation of our findings (Creswell, 2009; 2014).

We focused on early elementary teachers (both pre-service and in-service) and targeted current students majoring in early childhood education (state licensure for pre-K to grade 2) and elementary education (state licensure for grades 1–6), as well as alumni of these two programs from the past ten years. We obtained from the college's Education Department office email addresses for current students who had declared early childhood education or elementary education as one of their majors. We worked with the college's Alumni Relations Office to reach recent graduates of these two programs, and the office staff contacted the alumni with email addresses on file with their office.

After we were granted Institutional Review Board approval, we sent out a survey via Survey Monkey to currently enrolled undergraduate students ($n = 145$) majoring in early childhood and elementary education, out of which 27 responded. Additionally, the survey link was emailed to alumni from 2006 to 2015 ($n = 216$) who were either early childhood or elementary education majors, and 21 responded. Of the total 48 respondents, two were male.

The survey consisted of 50 questions—a combination of four-point Likert scale, multiple choice, and open-ended items. Survey questions were adapted from the Attitude Toward Math supplemental subtest of the *TOMA-3* (Brown, Cronin, & Bryant, 2012). Additional questions were taken from the math anxiety quiz (Chinn, 2012) and the abbreviated version of the mathematics anxiety rating scale (A-MARS; Alexander & Martray, 1989). Survey questions are listed in appendix A. All survey responses were anonymous. At the beginning of the 2016–2017 academic year, 33 early childhood and elementary freshman and transfer students were sent an abbreviated survey with 39 questions, out of which 10 responded. Of these 10 respondents, one was male.

Table 4.1. Participants

Level of Education	Total	Early Childhood	Elementary
Undergraduate	27	9	18
Alumni	21	9	12
Incoming Freshman/ Transfers	10	2	8
Total	58	20	38

At the end of the survey, participants had the option to volunteer for an interview with the undergraduate student. We anticipated that the students and alumni would feel more comfortable speaking with a peer rather than a professor. We also tried to maintain the anonymity of the participants since the lead author is a professor who teaches a required math methods course as well as courses in the graduate program at the college. The interviews were recorded with the participants' permission and then transcribed. The interview participants were assigned a number and their identity was not disclosed to the faculty member. Four undergraduate students and three alumni volunteered for a 30-minute semi-structured interview. Interview questions are listed at the end of the chapter.

In order to compare education majors to students in similar schools and programs, we interviewed five professors from three states (Massachusetts, Illinois, and Pennsylvania) who teach math methods or math content classes for elementary education majors at private colleges. We employed a *snowball*, or *chain*, *sampling* strategy to contact these professors. This approach identifies cases via people who know people in a certain environment (Creswell, 2007). In our case, the math professor at our research site referred us to his colleagues in professional organizations who taught math courses similar to the courses offered at our research site, and we followed up with them via email. The interviews were conducted using Skype and recorded with permission and transcribed.

We also conducted a document review of the current Massachusetts Guidelines for Mathematical Preparation of Elementary Teachers (Massachusetts Department of Education, 2007) and the Massachusetts Curriculum Framework for Mathematics (Massachusetts Department of Elementary and Secondary Education, 2011).

We analyzed the survey data using the statistical analysis software SPSS to determine descriptive statistics as well as frequency of responses. We coded the qualitative interview and short-answer survey data using key words and themes that emerged during the literature review and in response to our research questions, guided by our theoretical framework of social constructionism and a secondary growth mindset lens (including but not limited to confidence, attitude toward math, mindset, desire to grow, parent involvement, home environment, curriculum, and engagement).

The coding occurred as soon as the interviews were transcribed. We added additional themes to our codes as we listened to participants' interviews, and we constructed categories in areas where themes turned into patterns. As more interview data was collected, we went back to earlier interview transcripts and short-answer survey data to look for key words and references for new themes that emerged (e.g., teacher enthusiasm or passion, encouragement, support, challenge or rigor, and choice). Both members of the research team coded the data independently then met to discuss findings and triangulate the data and confirm emergent findings.

RESULTS

The results presented in the following sections reflect the themes and synthesis of the data.

Prior Experiences in Mathematics That Shape Beliefs about and Perceptions of Mathematics

Former math teachers' attitude, passion, enthusiasm, encouragement, and support influenced the participants' attitude toward, appreciation for, and confidence in mathematics. Teachers who encouraged students and focused on a growth mindset helped their students feel comfortable and less anxious about math. Nearly 80% of respondents ($n = 46$) remembered their teachers being excited about mathematics both at the elementary and secondary level. A majority of participants (67%; $n = 39$) reported that their teachers taught math in a way that was easy for them to understand.

Survey respondents cited specific examples of ways their teachers engaged them with mathematical content while providing a safe learning environment in which they could take risks. One respondent mentioned that a former teacher allowed students to collaborate when presented with challenging tasks:

> My best math teacher was rigorous but always set us up with tools we could use to solve a new type of problem . . . The class size was small so we all got a lot of attention and every Friday was a team challenge day where we would work together to solve problems like the ones we have been working on in class and logic problems. (Survey Respondent 46, July 21, 2016)

Another respondent reported that a former math teacher took extra time out of his day to help students make personal connections to statistics:

> My high school statistics math teacher really helped me to engage and understand what I was learning. He would provide opportunities for review sessions or sessions during lunch when students did not understand. He would use creative ways of teaching us how to use statistics in everyday living. (Survey Respondent 9, June 20, 2016)

One respondent recalled a high school math teacher who created a supportive learning environment:

> My high school calculus teacher was extremely patient, kind, and supportive. She was enthusiastic about math and made an effort to check in with every student. She showed videos and provided note packets that kept us engaged and interested. She was always available for extra help, and she showed that she

cared for us through her words and selfless actions. She would show us multiple representations of a math problem when we were confused and would not stop explaining it differently until we gained a greater understanding. (Survey Respondent 28, July 13, 2016)

On the other hand, teachers who did not allow students to ask questions or had poor classroom management skills led to students with a negative attitude toward math. In this sample, 13.6% of students and alumni ($n = 8$) remembered that their teachers were not enthusiastic about math. Additionally, nearly 35% of participants ($n = 20$) reported feeling anxious when they were asked to remember high school math skills.

Some respondents mentioned particular teachers as strongly affecting their negative perception of mathematics. According to one of the alumni, "Yeah, I guess . . . in high school, just because of . . . one algebra teacher, I had him two out of the four years, I decided not to take a math my senior year; he just really put me down" (Alumni 1, personal communication, July 8, 2016). Another respondent also mentioned their algebra teacher: "The teacher yelled at me for not knowing how to solve the problem and told me I was going to be screwed up for life because I didn't know how to do algebra. I wish he had taught me how to do it" (Survey Respondent 42, July 19, 2016).

One respondent mentioned their math classroom as a tense and threatening environment.

When I was in fourth grade, I had a terrible math teacher. She would constantly yell at us and make her students cry when they didn't get the right answers. The classroom environment she had was very judgmental and tense. She did not put time into helping us improve . . . She most definitely made me believe that I could not do math and that I would never be able to do math. I believe that if I had a different teacher in fourth grade, I might view math differently and actually enjoy it. (Survey Respondent 9, June 20, 2016)

From these participants' responses, it is clear that even a single teacher can affect a student's perception of math—sometimes permanently.

Other respondents spoke specifically about the way teachers engaged them in class, which helped them have a more positive view of math.

Good math teachers are . . . encouraging, motivating. I think they really have to enjoy math, and be really excited about what they are teaching—it really helps students to be more willing to want to understand and want to learn. Whereas if the teacher is not very excited or is just kind of dull, they're more likely not to have their students be focused or willing to learn. (College Student 1, personal communication, July 8, 2016)

Overall, participants enter pre-service teaching programs with a variety of background experiences in math content and math instruction, which can affect their view of math. Professors of math content and methods courses can learn more about students' background in mathematics by asking them to write a math autobiography at the beginning of the course (Professor 3, personal communication, July 21, 2016; Aguirre, Mayfield-Ingrim, & Martin, 2013). This assignment can help professors understand students' prior experiences and address negative attitudes toward math.

Beneficial Coursework and Classroom Experiences in Mathematics

Participants noted that their experience in math content coursework in college made them feel more confident to teach math. The math content courses allow pre-service teachers to develop a conceptual understanding of math concepts that they might have forgotten or never developed in their math experience. Of the respondents who had completed the math content courses, more than 72% ($n = 27$) saw math in an exciting light compared to 52% ($n = 11$) of those who have not taken the math courses.

According to open response answers in the survey, participants who had not yet taken math content courses reported feeling smart in math when they got answers right, completed problems quickly or better than their peers, and scored well on assessments (e.g., 90% or 100%). On the other hand, participants who had completed math content courses acknowledged more of a growth mindset approach to mathematics by remarking that they felt smart in math when they engaged with challenging problems, which fostered perseverance. This depth of thinking is in line with what mathematicians actually do in real life.

Participants mentioned specific college professors in a teacher-prep program as being instrumental in changing their perception of mathematics. One respondent reported that their math content course professor, who had taught the course for 37 years, created a safe environment in his classroom:

> The professor made it fun to learn about math. He would crack jokes and make the overwhelming feeling of math not exist. He brought a joy and perspective of math that I never experienced before. He would also thoroughly explain everything and re-explain if I did not understand. He also welcomed us to come to office hours or would make an appointment with us if we needed extra help. (Survey Respondent 18, July 12, 2016)

Another respondent also reflected on a professor's ability to minimize threats in his classroom: Our professor . . . changed the way I look at and do math . . . He encouraged me and spent so much time trying to help me when I was really struggling. He made sure his classroom was a comfortable and not a stressful environment" (Survey Respondent 9, June 20, 2016).

Through content and methods courses as well as contact with specific faculty members who create a safe environment for learning math, pre-service teachers may develop a more positive outlook toward mathematics.

DISCUSSION

Pre-service teachers may enter math methods and content courses with a traditional view of math and limited conceptual understanding, which can make them hesitant and anxious about teaching others math. This study sought to explore pre-service and novice teachers' beliefs about math at one private, liberal arts college. We explored courses and experiences that contributed to their beliefs during their undergraduate training. As we analyzed the data, we noticed five strands of math proficiency embedded in participants' responses.

Teachers who have a strong conceptual understanding, procedural fluency, strategic competence, adaptive reasoning, and a productive disposition toward math themselves are able to model and develop these strands in their students as well (National Research Council, 2001). These proficiencies have now evolved into the eight math practice standards as part of the common core state standards as practices that math educators should seek to develop in their students (Common Core State Standards Initiative, 2016). Primarily, teachers of math need to be proficient in mathematics themselves if they are to foster math proficiency in their students.

Conceptual Understanding

Teachers with a strong conceptual understanding of math comprehend the underlying concepts and relationships in mathematical problems. These teachers have a deep understanding of why the mathematical steps work in a problem and can show their thinking using several types of materials. These teachers help students use their quantitative reasoning skills by facilitating mathematical discussion about what is happening in a problem.

Participants practice justifying their answers and explaining their reasoning in their college math content class. The focus of the math content class is not solely on finding the right answer but on understanding the processes behind the mathematical calculations and why they work.

Pre-service teachers must display conceptual understanding by verbally justifying their solutions and approaches. For them to feel comfortable doing this in class, their professors should provide time for students to explain their thinking in class to their peers. Participants stated that their past teachers encouraged students and spent extra time outside class helping them understand math concepts. Students must have a willingness to learn and understand, and a safe and supportive learning environment can foster a positive outlook toward math.

Procedural Fluency

Teachers and students work to be fluent in mathematics. They are flexible, accurate, and efficient in carrying out appropriate mathematical procedures. They know the patterns and underlying structure of mathematical problems. Instead of relying on standard methods or procedures, high-achieving students in math are flexible with numbers and often use strategies to compute mentally. Thinking deeply and carefully about math does not necessarily mean that you are the fastest—and procedural fluency involves more than speed. Students who are truly fluent in mathematics are able to choose an appropriate strategy based on the numbers and situation of a particular problem. This may or may not mean using a traditional standard algorithm (Van de Walle, Karp, & Bay-Williams, 2016).

According to Professor 5, math is all about seeing connections. "Math isn't primarily about speed, it's much more about taking the time to think deeply" (personal communication, July 26, 2016). Students feel less anxious about math if the emphasis is on flexibility and accuracy rather than just speed.

Strategic Competence

Math teachers and learners are strategic. They have the ability to represent and solve problems in multiple ways. Teachers and students of math can use a variety of strategies to solve a problem and are encouraged to show several solution pathways. Teachers give students time to form their arguments and explain their thinking to their peers.

In college math content courses, professors have pre-service teachers consider ways to represent the math concepts in multiple ways. Professor 3 explained how she prepares her pre-service teachers to practice this skill before they enter the classroom:

> I'll just give them a problem and say a student solved it this way, does it work and why? Rather than me saying, this is how it works, just memorize this and do it just like I do it. You know, try to have them analyze it first and say, oh this is a viable method. Because what's going to happen in the classroom is there's going to be students that find all different methods to solving problems, and you as the teacher have to be able to analyze and say, that is viable, and not just say,

no that's wrong do it my way, you know? Because what does that do to that child? (personal communication, July 21, 2016)

Adaptive Reasoning

Math teachers and students have the ability to reflect on, justify, and explain the way they solved a problem. They can communicate why they know their answer makes sense. Teachers use mathematical facilitation skills to ask students questions and prompt their students to ask questions of each other.

One professor models this for her pre-service teachers in a math content course since she has students collaborate with their peers and justify their reasoning when they solve problems. Professor 3 has students solve problems together and explain their reasoning to their peers, which lowers their anxiety (personal communication, July 21, 2016). In math content and methods courses, professors model adaptive reasoning skills for pre-service teachers to incorporate in their future classrooms.

Productive Disposition

Math teachers and students have the ability to view math as meaningful and worthwhile, and they believe it is something you can do and learn to do. With a productive disposition toward math, teachers are passionate about the subject and seek to engage their students with math content. A productive disposition toward math corresponds to a growth mindset in math, a belief that with hard work, effort, and perseverance you can learn and achieve in math (Boaler, 2016). Participants who had taken the math content course seemed to move away from a fixed mindset and become more aligned with a growth mindset approach to math (see Table 4.2). Over 80% of participants ($n = 30$) who had taken a math content course responded negatively to a fixed mindset statement about math compared to 47.6% of participants ($n = 10$) who had not yet taken a math content course.

Table 4.2. Response: I've Never Been Good At Math and I Don't Think I Will Ever Be Good At It

Perceptions	Participants Who Have Taken a Math Content Course		Participants Who Have Not Taken Math Content Course	
	n	%	n	%
No, definitely!	21	56.8	4	19.0
Closer to no	9	24.3	6	28.6
Closer to yes	3	8.1	8	38.1
Yes, definitely!	1	2.7	2	9.5
N/A	3	8.1	1	4.8

Former teachers seemed to play an important factor in participants' attitudes toward math. Participants had a variety of previous experiences in math. According to the open response questions in the survey, participants compared the characteristics of their best math teachers with those of their worst math teachers (see Table 4.3).

Table 4.3.	Comparison of Strong Math Teachers and Poor Math Teachers

Strong Math Teachers	Poor Math Teachers
Accepted flexible methods of solving a problem.	Showed or accepted one way to complete a problem.
Facilitated learning.	Gave knowledge, lectured.
Made connections to real-world examples.	Focused on textbook problems.
Encouraged student participation.	Didn't encourage discussion or provide an opportunity to ask questions.
Created an accepting and supportive environment.	Gave up helping or assisting.
Was passionate about the subject.	Was dull, not particularly interesting.
Varied the pace of work.	Rushed instruction.

Participants focused on the qualities of their former math teachers that centered on developing a productive disposition toward math and facilitated their engagement with math content. Math teachers can help students develop a positive attitude toward math when students have a positive learning experience in classrooms that encourage participation through discussion facilitation and connection to real-life examples.

Overall, pre-service teachers and current teachers represented a spectrum of comfort levels in math due to previous math experiences. We acknowledge that pre-service teachers enter the program with a variety of previous experiences, which influence their perceptions of mathematics. However, students have a positive experience in math content courses at the college level in the program.

CONCLUSIONS AND RECOMMENDATIONS

Educators must first persist and persevere in their own math skills if they are to model these skills for their elementary students as evidenced by Math Practice Standard 1: Make sense of problems and persevere in solving them (Common Core State Standards Initiative, 2016). When teachers are empowered in mathematical thinking, their students will also be empowered (Ma, 1999).

According to Slavin, Lake, & Groth (2010), the most important factor in students' mathematical achievement is teaching practices that encourage student interaction (cooperative learning and motivation). Strong academic standards and quality curriculum are important to mathematics instruction, yet skilled teachers who are confident in their mathematical ability are needed to engage and support students in order to learn mathematics (Ball, Hill, & Bass, 2005).

In our pre-service teacher education coursework, students can focus even more on self-examination to understand and challenge their beliefs, attitudes, and actions toward math as well as the way these beliefs will affect their teaching of math and their impressions of their elementary students. Professors can address math anxiety and past math trauma through a math autobiography and revisit it throughout the course. Students can ask themselves, What stereotypes do I hold about math? What is my mindset toward math? What do I believe about my potential as a strong mathematical learner? What do I believe about my students' potential? What messages am I sending to my students about math? (Aguirre et al., 2013).

In higher education, professors can address pre-service teachers as "mathematicians" as they engage in mathematical inquiry and thinking in math content and math methods coursework. Professors of math content and math methods courses can also create safe spaces in which pre-service teachers can discuss their strategies and are rewarded for effort and perseverance. Professors can ask students to discuss their solution pathways with a partner ("Think-pair-share") before sharing with the whole group to enable students to feel more confident sharing their responses. Pre-service teachers must experience a growth mindset if they are going to pass it on to their prospective students.

The researchers note limitations including the use of a case-study approach at a small, private, liberal arts college. Also, 58 surveys were received, which is a small sample size. More research at other types of higher education institutions would be beneficial so that the results might be more usefully generalized.

SURVEY QUESTIONS

1. I have read the informed consent form and agree to participate in this study.
   a. Yes
   b. No (if you choose this option, you will be unable to complete this survey a second time)
2. Are you/Did you major in early childhood education or elementary education?
   a. Early Childhood Education

   b. Elementary Education
3. I am a current undergraduate student.
   a. Yes
   b. No

Undergrad Questions

4. Have you taken MATH 205—Concepts of Math I?
   a. Yes
   b. No/Not yet
5. Have you taken MATH 206—Concepts of Math II?
   a. Yes
   b. No/Not Yet
6. How would you rate Concepts of Math I and/or II in comparison to other courses you have taken at the College?
   a. Easier than other courses
   b. Somewhat easier than other courses
   c. The same as other courses
   d. Slightly more difficult than other courses
   e. The most difficult course I took
7. How did you feel about Concepts of Math? (if applicable)
   Open response
8. Have you taken EDU 270—Math Methods?
   a. Yes
   b. No/Not yet

Alumni Questions

9. I have already graduated from the College.
   a. Yes
   b. No
10. I am currently teaching in a classroom.
    a. with my initial license
    b. with my professional license
11. How would you rate Concepts of Math I and/or II in comparison to other courses you have taken at the College?
    a. Easier than other courses
    b. Somewhat easier
    c. The same as
    d. Slightly more difficult
    e. The most difficult

12. How did you feel when you took Concepts of Math I or II—MATH 205 and 206?
Open Response

Alumni and Undergrad Questions

13. Which MTEL have you taken and passed that required mathematics?
    a. Elementary Math
    b. General Curriculum—Mathematics
    c. Early Childhood
    d. I have not yet taken any of these MTELs.
    e. I have taken and not yet passed any of these MTELs.
14. How many times did you attempt to pass one of the MTELs in the previous question?
    a. I passed on the first attempt.
    b. I passed on the second attempt.
    c. More than two attempts
15. Be honest in describing how you feel:
    a. It's fun to work math problems.
    b. You should take math every year.
    c. If I could have skipped just one class it would have been math.
    d. I'm better at math than most of my peers.
    e. Math is interesting and exciting.
        1. Yes, definitely! (Agree)
        2. Closer to yes
        3. Closer to no
        4. No, definitely! (disagree)
16. When do you feel smart in math?
Open Response
17. Be honest in describing how you feel:
    a. Math tests were usually easy for me.
    b. I'd rather do math than any other kind of homework.
    c. I liked every other subject in school better than math.
    d. Someone who likes math is usually weird.
    e. I use a lot of math outside of school.
        1. Yes, definitely! (Agree)
        2. Closer to yes
        3. Closer to no
        4. No, definitely! (disagree)
18. How do you feel during a math test?
Open Response

19. Be honest in describing how you feel:
    a. I've always liked math.
    b. I've never been good at math, and I don't think I will ever be good at it. There are just some people that are good at math, and I'm not one of them.
    c. My teachers made math easy for me to understand.
    d. My parents/guardians encouraged me to engage in problem solving activities and games/puzzles when I was young.
    e. My friends like math more than I do.
       1. Yes, definitely! (Agree)
       2. Closer to yes
       3. Closer to no
       4. No, definitely! (disagree)
20. Write about the best math teacher you've had.
    Open Response
21. Write about the worst math teacher you've had.
    Open Response
22. Be honest in describing how you feel:
    a. My parents told me that it was okay if I wasn't good at math, since they weren't good at math.
    b. My elementary teachers showed interest and excitement toward math.
    c. My middle and high school teachers showed interest and excitement toward math.
       1. Yes, definitely! (Agree)
       2. Closer to yes
       3. Closer to no
       4. No, definitely! (disagree)
23. What is difficult about teaching math? (Skip, if you have no experience in this area.)
    Open Response
24. What do you enjoy about teaching math? (Skip, if you have no experience in this area.)
    Open Response
25. The items below are about math and your feelings when you have to do each one of these things or have had to do them in the past. Consider each item and then decide how anxious that situation makes you feel.
    a. Working out the tip for the waiter in a restaurant
    b. Checking the cost of your shopping
    c. Working out the 20% off in a sale
    d. Studying for a math test
    e. Working out the cost of a vacation

f. Thinking about an upcoming math test one day before
g. Having to recall a math fact quickly (such as 6 x 9)
h. Walking into math class
i. Changing the quantities in a recipe for four when cooking for six people
j. Remembering your math skills from high school
k. Splitting the check or bill with friends at a restaurant
l. Looking at your final grade in a math class
m. Signing up for a math course
n. Watching a teacher work on an algebraic equation on the blackboard
   1. Never anxious
   2. Sometimes anxious
   3. Often anxious
   4. Always anxious
26. Sex: Male or Female?
   a. Male
   b. Female

INTERVIEW QUESTIONS FOR CURRENT STUDENTS AND ALUMNI

1. Describe your math achievement in middle school and high school. For instance, were you tracked by math ability? Which track did you enter and what was your experience?
2. How do you think you were perceived by your math teachers? Perhaps due to the track you were on, or in general.
3. Describe an "A-ha!" moment (a moment of insight when you did math well) in a math content class or math methods class.
4. What surprised you in the transition to college math?
5. What interactions have you had with a math coach or math specialist? What is their role at your school?
6. What qualities make a good math teacher?
7. How do you help students experience success in math? How do you define success in a math class?
8. How do you generate motivation and engagement among your students in mathematics?
9. How do you individualize instruction for students in mathematics?
10. How do you get students interested in the problem-solving process?
11. How do you evaluate student math progress other than with testing?

12. Do you give math homework? If so, how much? What is a typical assignment like? What have you learned over the years about giving math homework?
13. What else you would like to tell me about your experiences with math, or teaching math?

REFERENCES

Alexander, L., & Martray, C. (1989). The development of an abbreviated version of the Mathematics Anxiety Rating Scale. *Measurement and Evaluation in Counseling and Development, 22*(3), 143–150.

Aguirre, J. M., Mayfield-Ingram, K., & Martin, D. (2013). *The impact of identity in K–8 mathematics learning and teaching: Rethinking equity-based practices.* Reston, VA: National Council of Teachers of Mathematics.

Ball, D. L., Hill, H. C., & Bass, H. (2005). Knowing mathematics for teaching: Who knows mathematics well enough to teach third grade, and how can we decide? *American Educator, 29*(1), 14–17, 20–22, 43–46.

Beijaard, D., Meijer, P. C., & Verloop, N. (2004). Reconsidering research on teachers' professional identity. *Teaching and Teacher Education, 20*, 107–128. doi:10.1016/j.tate.2003.07.001

Beilock, S. L., Gunderson, E. A., Ramirez, G., & Levine, S. C. (2010). Female teachers' math anxiety affects girls' math achievement. *Proceedings of the National Academy of Sciences, 107*(5), 1860–1863. doi:10.1073/pnas.0910967107

Boaler, J. (2015). *What's math got to do with it? How teachers and parents can transform mathematics learning and inspire success.* New York, NY: Penguin.

Boaler, J. (2016). *Mathematical mindsets: Unleashing students' potential through creative math, inspiring messages, and innovative teaching.* San Francisco, CA: Jossey-Bass.

Brown, V., Cronin, M., & Bryant, D. (2012). *The test of mathematical abilities, 3rd ed. (TOMA-3).* Austin, TX: Pro-Ed.

Burr, V. (1995). *An introduction to social constructionism.* New York, NY: Routledge.

Busch, B. (2017). *Growth mindset: Practical tips you may not have tried yet.* Retrieved from www.theguardian.com/teacher-network/2017/jan/09/growth-mindset-practical-strategies-classroom

CAST. (2016). *Top 5 UDL tips for reducing stereotype threats.* Retrieved from https://gallery.mailchimp.com/89e11c7455f4cb757154eb608/files/cast_5_stereotype_threat.pdf

Chinn, S. (2010). *Maths anxiety quiz.* www.stevechinn.co.uk/maths-quiz.html

Common Core State Standards Initiative. (2016). *Standards for mathematical practice.* www.corestandards.org/Math/Practice/

Creswell, J. W. (2009). *Research Design: Qualitative, quantitative, and mixed methods approaches, 3rd ed.* Thousand Oaks, CA: Sage Publications.

Creswell, J. W. (2014). *Research Design: Qualitative, quantitative, and mixed methods approaches, 4th ed.* Thousand Oaks, CA: Sage Publications.

Drake, C., Spillane, J. P., & Hufferd-Ackles, K. (2001). Storied identities: Teacher learning and subject-matter context. *Journal of Curriculum Studies, 33*(1), 1–23.

Dweck, C. S. (2006). *Mindset: The new psychology of success.* New York, NY: Random House.

Gergen, K. (1985). The social constructionist movement in modern psychology. *American Psychologist, 40*(3), 266–275.

Gray, E. M., & Tall, D. O. (1994). Duality, ambiguity, and flexibility: A "proceptual" view of simple arithmetic. *Journal for Research in Mathematics Education, 25*(2), 116–140.

Koestler, C. (2012). Beyond apples, puppy dogs, and ice cream: Preparing teachers to teach mathematics for equity and social justice. In A. A. Wager & D. W. Stinson (Eds.), *Teaching mathematics for social justice: Conversations with educators* (pp. 81–97). Reston, VA: National Council of Teachers of Mathematics.

Ma, L. (1999). *Knowing and teaching elementary mathematics: Teachers' understanding of fundamental mathematics in China and the United States.* Mahwah, NJ: Lawrence Erlbaum Associates.

Massachusetts Department of Education. (2007). *Massachusetts guidelines for mathematical preparation of elementary teachers.* Retrieved from www.doe.mass .edu/mtel/mathguidance.pdf

Massachusetts Department of Elementary and Secondary Education. (2011). *Massachusetts curriculum framework for mathematics.* Retrieved from www.doe.mass .edu/frameworks/math/0311.pdf

McCray, J. (2016). In response to "the wrong way to teach math." Retrieved from http:// earlymath.erikson.edu/response-wrong-way-to-teach-math/ and www.preschool mathactivities.org/in-response-to-the-wrong-way-to-teach-math/

National Research Council. (2001). *Adding it up: Helping children learn mathematics.* J. Kilpatrick, J. Swafford, & B. Findell (Eds.). Washington, DC: National Academies Press.

Remillard, J. T. (2000). Prerequisites for learning to teach mathematics in multicultural contexts. In W. G. Secada (Ed.), *Changing the faces of mathematics: Perspectives on multiculturalism and gender equity* (pp. 125–136). Reston, VA: National Council of Teachers of Mathematics.

Slavin, R., Lake, C., & Groff, C. (2010). What works in teaching math? Retrieved from www.bestevidence.org/word/math_Jan_05_2010_guide.pdf

Spillane, J. P. (2000) Constructing ambitious pedagogy in the fifth grade: The mathematics and literacy divide. *Elementary School Journal, 100*(4), 307–330.

Steele, C. (2010). *Whistling Vivaldi: How stereotypes affect us and what we can do.* New York, NY: W. W. Norton.

Turner, S. (2013). *Teaching primary mathematics.* London: Sage Publications.

Van de Walle, J. A., Karp, K., & Bay-Williams, J. M. (2016). *Elementary and middle school mathematics: Teaching developmentally* (9th ed.). Boston, MA: Pearson.

Chapter 5

A Teacher's Perspective of Quantitative Literacy in Middle School Mathematics

Heather Crawford-Ferre and Diana L. Moss

INTRODUCTION

Eric Gutstein in his book *Reading and Writing the World with Mathematics: Toward a Pedagogy for Social Justice* (2006a) wrote,

> I believe that mathematics is important because math is part of our life, perhaps the most important part of our life. There are numbers when we wake up, there are numbers when we eat, there are numbers when we work, there are numbers wherever we go, there are numbers when we go to sleep, everywhere in life there are numbers. (p. 343)

The authors of the study reported in this chapter took direction from Gutstein's words in framing the study. The complexity of mathematics is most important in the preparation of mathematics teachers and in the teaching of mathematics in the classroom with each new generation of students.

The purpose of this study was to determine the effectiveness of a diagnostic interview assignment for helping pre-service teachers (1) learn how to conduct a diagnostic interview successfully and (2) learn how to describe student understanding of a mathematical topic (subtraction) based on evidence from the interview and grounded in research literature. The study sought to determine how well course readings and instructional activities prepared pre-service teachers to analyze the results of the diagnostic interviews. We performed a backward analysis of pre-service teachers' diagnostic interview papers in the context of the course content to determine how pre-service teachers interpreted students' mathematical thinking and how they reflected on, and discussed, future implications for teaching based on their analyses of student thinking.

QUANTITATIVE LITERACY

Studies indicate that students are not as quantitatively literate as they ought to be (see Huang [2004]; and Wilkins [2000]). For example, Huang's study of 48 fourth-grade children's performance on and perception of problem difficulty on word problems with familiar and unfamiliar contexts showed that "more than half the children did not identify the similarity in problem-solving approaches between problem settings and real shopping" (p. 278).

Further, Wilkins (2000) found that school completers in the United States were less capable of applying their content knowledge to everyday situations than school completers in other countries who participated in the Third International Mathematics and Science Study. This is especially problematic given that quantitative literacy (QL) is growing in importance within this increasingly data-dense world (Wilder, 2010). In addition to its relation to preparation for employment, QL is linked to informed citizenship and life quality through areas such as health, education, and finance (Wiest, Higgins, & Frost, 2007). Students who are not quantitatively literate are ill prepared to participate in society and to make effective everyday decisions.

QUANTITATIVE LITERACY, TEACHER KNOWLEDGE, AND IMPLEMENTATION

QL is different than typical school mathematics. School mathematics tends to focus on abstract, procedural knowledge with little problem-based, investigative thinking. QL, in contrast, involves the use of mathematics in the context of real-world scenarios. QL is supported by the National Council of Teachers of Mathematics (2000) and the National Governors Association Center for Best Practices, Council of Chief State School Officers (2009), who have established national standards that advocate use of "real-world" mathematics.

However, real-world mathematics is typically implemented in the form of story problems that involve hypothetical stories rather than authentic tasks (Wiest, Higgins, & Frost, 2007). For true implementation of QL, students must apply fundamental mathematics to genuine, real-world, interdisciplinary situations.

Unfortunately, many teachers do not fully understand how mathematics relates to the real world (Garii & Okumu, 2008) and "generally do not teach the types of knowledge, skills, and dispositions that QL requires, thus shortchanging students in their readiness for real-world demands" (Wiest, Higgins, & Frost, 2007, p. 50). Pre-service teachers tend to exclude real-world information, such as tax in solving shopping problems, and thus preference non-realistic answers to real-world problems (Verschaffel, Greer, & DeCorte, 2000), which indicates that pre-service teachers tend not to approach problems from a QL standpoint.

Further, Gutstein (2006) found than many teachers do not appreciate the practical utility of the mathematics they teach. For example, in a survey of 62 teachers, Gainsburg (2008) found that teachers reported giving real-world examples rather than engaging students in tasks with real-world data. Gainsburg recommends professional development to address teachers' lack of appropriate QL implementation.

In England, teacher-training programs in numeracy (QL) have been implemented. Trainees report a wider view of mathematics, demonstrating that these courses accomplished the goal of expanding teachers' view of mathematics teaching (Loo, 2006). This highlights the importance of teacher education in helping teachers to see the real-world applications of mathematics.

Garii and Okumu (2008) assert that one reason teachers may limit or omit QL topics is lack of teacher training in embedded mathematics topics, which situate mathematics content in real-world scenarios. They claim mathematics education for pre-service teachers should include exploration of QL to increase implementation of QL instruction in K–12 education. Howe (2003) asserts that weak teacher QL means that better pre-service and in-service teacher education might yield better QL achievement for students. Given the connections between professional development and implementation it is imperative to investigate professional development formats to support teachers.

ONLINE EDUCATION

Online education has developed as an alternative to traditional professional development for teachers. This form of distance education potentially mitigates challenges caused by geographic and time constraints. Additionally, online education provides an alternative to recruit teachers for continuing education who prefer the flexibility and instructional delivery of online instruction (Crawford-Ferre & Wiest, 2012). As a result of these factors, Major (2010) found that nearly 100% of public institutions of higher education report online instruction as a critical part of their long-term plans.

Further, some institutions of higher education (e.g., the Stanford Center for Professional Development) offer non-credit-bearing online professional development opportunities. In many states and school districts, online professional development courses are honored as evidence of continuing education for license renewal and salary advancement. Further, Smith (2012) found that perceived "ease of use, perceived usefulness and social presence" significantly predicted teachers' preference for e-learning for professional development.

PURPOSE OF THE STUDY

The purpose of this study was to investigate how a middle school teacher implemented QL in her mathematics classroom after participating in an online, three-credit, graduate-level, professional development course titled "Critical Numeracy Across the Curriculum."

The following research questions guided this study:

1. How does a middle school teacher perceive her implementation of QL after online professional development?
2. How does the teacher implement QL instruction in her middle school classroom after online professional development?

CONTEXT AND METHODS

QL and the preparation of pre-service teachers in mathematics in this study focused on diagnostic interviews and how well course readings and instructional activities benefited the preparation of math teachers. The context and methods used in the study are addressed in the sections that follow.

Research Design

The researcher employed grounded theory for data analysis (Glaser & Strauss, 1967) and simultaneously collected and analyzed data (Lichtman, 2011). The analysis was a part of the research design with the coding of the first data set serving as a foundation for future data collection and analysis (Corbin & Strauss, 2008). The researcher adjusted additional data collected based on analysis of the first data set (Corbin & Strauss, 2008) by basing the interview questions on analysis of the initial data.

The researcher proceeded through the coding process by analyzing the data line by line and paragraph by paragraph and then coding the deconstructed fragments (Lichtman, 2011). Codes were compared, renamed, added, or deleted as the researcher constantly compared them (Corbin & Strauss, 2008).

University Site and Course Description

A graduate course on QL, Critical Numeracy Across the Curriculum, was offered at a land grant university located in the western United States. Eighteen master's and doctoral students were enrolled in the course, including kindergarten to post-secondary teachers and full-time graduate students who were not currently teaching. The course was offered online in the WebCampus format and included doing assigned readings, discussing designated topics using an asynchronous electronic forum, writing a QL lesson, providing feedback on other students' QL lessons, writing scholarly essays, and taking a proctored exam.

Focus Study Participant and Her Classroom Context

The teacher selected for this research was chosen because she was the only general education middle school mathematics teacher who was enrolled in the QL course. She had taught mathematics for four years, was in her second year teaching middle school, and was in her first year teaching at her current school.

The teacher had earned a bachelor's degree in discrete mathematics, a master's degree in applied mathematics, and a secondary teaching license program, and she was currently enrolled in a doctoral program in mathematics education. The teacher taught at a small parochial school with an enrollment of 288 students, including 35 eighth graders. The teacher's math class took place three days each week for a total of five hours. Twenty-two of the eighth graders were enrolled in algebra and were taught by this teacher. These students were 12 to 13 years old and included 12 female and 10 male students. Most were from middle- to upper-class families, and all were White.

Data Sources and Data Collection

Initially, the researcher conducted an interview with the teacher, asking her to describe QL. Additionally, the participant was asked to identify ways she implements QL in the classroom. The researcher then observed the classroom instruction three days a week for two months. She conducted observations as described by Florio-Ruane (1999), mapping the classroom, making a classroom log, and collecting field notes. After the observations, the researcher conducted a second interview to ask questions prompted by the observation. The following questions were asked:

- What discourages/inhibits using QL in the mathematics classroom?
- What encourages/prompts using QL in the mathematics classroom?
- Has there been any change in the QL you teach in your mathematics class-room? If so, what led to that change?
- Why do you use QL in your classroom?
- How do you implement QL in your classroom?
- Where do you get your QL lesson plan ideas?
- What do you perceive as a benefit, if any, to using QL?
- What do you perceive as a drawback, if any, to using QL?

Finally, the researcher collected the following artifacts: lesson plans for an eight-week quarter, and copies of the assignments given to students for an eight-week quarter.

Data Analysis Procedures

After each week of observation and each interview, the researcher wrote data-analysis memos (Maxwell, 2004). This created eight memos based on observations reflecting on initial analyses. Both interviews were transcribed by the researcher, who wrote the data-analysis memos (Maxwell, 2004) as she kept in mind the research questions while transcribing. The first interview was analyzed prior to conducting the second interview (Lichtman, 2011). Next the researcher systematically studied all field notes taken during classroom observations (Florio-Ruane, 1999).

To analyze the data set, the researcher used grounded theory by employing a constant comparison technique as described by Cohen, Manion, and Morrison (2007). Using constant comparison, the researcher constructed conceptual categories based on multiple re-readings and re-coding of the data. As classification of data continued, codes were renamed, added, combined, or removed to arrive at final categories (Cohen et al., 2007). "This 'checking back' is a method of confirming or disconfirming that ensured that the categories were grounded in the theory rather than 'flights of fancy' or pet ideas" (Lichtman, 2011, p. 63). The artifacts collected, including lesson plans and assignments, were analyzed for consistency with the observations of teaching and the teacher's reports of QL implementation in the classroom.

RESULTS

The results are categorized into the following themes: authentic experience, connection to mathematics standards, teacher knowledge, and student engagement. Within each section, the teacher's self-reported perceptions are discussed first and then addressed within the context of the classroom observations.

Authentic Application of Mathematics Concepts

The teacher reported a goal of using mathematics authentically as her primary reason for implementing QL in her teaching. She stated that it was important that her students know how to research real data, represent it using graphs and charts, and analyze it. Further, she indicated that it was important for her students to know mathematics rules and have the ability to apply them in their daily mathematics encounters, saying, "I use QL in my class because it is important for students to be able to apply their mathematical skills to the real world. Applying their knowledge to solve a problem in a different context involves a deeper level of thought."

This perception of authentic instruction was supported frequently in the participant's teaching. The teacher's lessons, homework assignments, and assessments could be categorized into four levels:

1. Students research authentic data to analyze and respond to a question(s).
2. Students are provided authentic data to analyze and respond to a question(s).
3. Students use hypothetical data to respond to a question(s) set in a real-world context.
4. Students use no QL.

The majority of the lessons fell into categories 1–3. Lessons that were not based on QL occurred only on days that were dedicated to review for a state assessment and on one day when technological problems prevented the planned lesson from being implemented.

The teacher had students use laptop computers to research authentic data and provided instruction on identifying reputable Web sources. Many lessons required students to find their own data to analyze. For example, in one lesson, students were asked to apply their knowledge of graphing systems of equations to research the cost of two gym memberships, create equations for both, graph the system of equations, and write a conclusion regarding which membership they would choose and why. When the students were not researching their own data, they were frequently analyzing authentic data provided to them such as data on the average unemployment rates and salaries of individuals with different levels of educational attainment.

The teacher's classroom actions reinforced her reported authentic use of mathematics. When students asked for information, she told them to look it up themselves. For example, when the students were working on a project generating algebraic sentences to describe the cost of owning a dog for a year, a student asked how to convert English pounds to dollars. In response the teacher asked the student to look up the information online and then ask for help if there was still a question.

The assessments and homework assignments, however, tell a different story. No assessment was given during this study that evaluated any QL skill or task. Further, the only required written QL homework assignments were incomplete classwork that students were asked to complete as homework. Typical homework included worksheets of algebra problems. The students also had electronic homework on a class webpage. Approximately one-quarter of the online assignments had a QL component, requiring the students to discuss or comment on the QL activities that were completed in class.

Connection to Math Standards

The teacher reported that curricular standards discourage her use of QL in the classroom, saying, "QL is not a part of the regular curriculum in eighth grade. There is not enough time to cover the mandated material as well as teaching QL." She elaborated that she thought there should be a separate QL course so that students would have time to apply their algebraic knowledge. Despite self-reported struggles with teaching QL due to its lack of alignment with the adopted standards, her teaching demonstrated use of QL in conjunction with algebra content.

Activities included writing number sentences based on the cost of owning pets, writing systems of equations based on gym-membership fees, and working with conversions and fractions based on food recipes. Not all QL lessons, however, were aligned with grade-level standards. One day was spent teaching an introduction to graph theory set in a real-world scenario and another day to having students compare the nutritional information from places they eat with the U.S. Department of Agriculture's nutritional guidelines.

Teacher Knowledge

The teacher reported that taking the course in QL changed her instruction. She said, "Before taking the QL course, I never taught QL in algebra. After the course, I taught QL every week. I learned what QL was and how to write lessons incorporating QL. I also learned why QL is so important." She also reported using lesson plan ideas directly from the course.

In the classroom, the teacher was explicit in using QL with her students. She explained what QL was and when they were using it. She acknowledged when activities were set in a real-world context but were not true QL. For example, when working with graph theory in the context of planning the most effective road trip, she told the class that this activity was not true QL because it was not using authentic data.

Student Engagement

The teacher reported that she used QL as a way to motivate her students to learn and felt they were more engaged when working with mathematics in the real world. She explained, "The main benefit of using QL is that students are more interested in the problems and like solving them since they apply to their lives." This perception was supported by classroom observations.

When students were off task during a QL project, she told the class that if their behavior continued, she would be forced to teach "regular" algebra instead, reinforcing her perception that her students were motivated by QL. During observations, she regularly commented on how engaged students were. For example, when doing a Webquest on supply and demand, she commented that she was impressed at how engaged the students were with the task. On a second occasion when the students were working on a QL task, she asked the principal to come and observe the level of student engagement.

DISCUSSION

These results indicate that taking a teacher education course in QL increased this teacher's QL instruction in her middle school mathematics teaching. This supports the call for QL instruction in college (Chacko, 2007; Steenken, 2007) and, in particular, teacher education (Garii & Okumu, 2008; Howe, 2003).

Further, the teacher asserted that the course was practical in terms of QL implementation, stating that she used complete lesson plans and lesson plan ideas directly from the course. This might help inform design of an effective teacher education course in QL. The role of a specific course in QL, rather than increased requirements in mathematics, is supported by Gutstein's (2006) plea that teachers need to learn how to read and write the world with mathematics in addition to possessing both disciplinary mathematics content knowledge and pedagogical content knowledge.

Although the course provided the teacher who participated in this study with the resources to implement QL instruction in her classroom, she did not conduct QL assessments. When asked about this in a follow-up interview, the teacher indicated that she was not sure how to assess this knowledge and did not feel she needed to assess QL since it was not a required standard.

Lack of assessment in QL is noted by Pugalee, Hartman, and Forrester (2008), who state, "Despite the importance of such skills, assessment of quantitative literacy has not been a focus in education and there is a lack of assessment tools that specifically address these skills" (p. 35). There is both a need for research and development in the area of formalized QL assessment and a need for explicit instruction in informal classroom QL assessment in teacher education courses.

The tie to standards also needs explicit instruction in QL education courses. The teacher in this study repeatedly reported her perception that QL was something extra—an enrichment as well as a time strain because it was not an eighth-grade standard. QL, however, is supported by the National Council of Teachers of Mathematics (2000) and the National Governors Association Center for Best Practices, Council of Chief State School Officers (2009), which advocate the use of "real-world" mathematics.

The view that QL is outside required standards is not held only by this teacher. Steen (2001) reports that despite the cooperation of the American Statistical Association and the National Council of Teachers of Mathematics, mathematics instruction has remained focused on school-based knowledge rather than real-world knowledge.

CONCLUSION

Despite not seeing the connection between established curricular standards and her QL instruction, the teacher in this study remained dedicated to infusing QL into her mathematics instruction, repeatedly reporting that she believed the instruction engaged her students. As noted above, she even requested that the principal observe student engagement during her use of QL in the classroom. Increased interest and participation in mathematics during QL tasks has been found in both K–12 and higher education instruction (see Briggs, Sullivan, & Handelsman [2004]; and McNamer [2009]).

This teacher attributed much of her student engagement to the technology she used in the classroom, including having the students use laptops and wireless Internet access to research their own data to solve problems, and looking at mathematics applications in everyday technology—such as the theorems used to create Mp3 technology—and representing the data graphically. She reported that she did not know if it was the technology or the actual QL task that led to the increased student engagement.

This perception of a dichotomy between QL and technology discounts the broadness of QL. Steen (2001) explicitly includes computer skills as an expression of QL, and he notes that numeracy is a part of every aspect of life.

In order to increase QL instruction in K–12 classrooms as well as teacher education, teachers must have a full understanding of what qualitative literacy encompasses.

REFERENCES

Beckett, G. H., Amaro-Jimenez, C., & Beckett, K. S. (2010). Students' use of asynchronous discussions for academic discourse socialization. *Distance Education, 31*(3), 315–335.

Briggs, W., Sullivan, N., & Handelsman, M. M. (2004). Student engagement in a quantitative literacy course. *AMATYC Review, 26*(1), 18–28.

Chacko, I. (2007). Real-world problems: Teachers' evaluation of pupils' solutions. *Studies in Educational Evaluation, 33*, 338–354.

Cohen, L., Manion, L., & Morrison, K. (2007). *Research methods in education* (6th ed.). New York, NY: Routledge.

Corbin, J., & Strauss, A. (2008). *Basics of qualitative research* (3rd ed.). Los Angeles, CA: Sage Publications.

Crawford-Ferre, H. G., & Wiest, L. R. (2012). Effective online instruction in higher education. *Quarterly Review of Distance Education, 13*(1), 11.

Florio-Ruane, S. (1999). Revisiting fieldwork in pre-service teachers learning: Creating your own case studies. In M. A. Lundeberg, B. B. Levin, & H. L. Harrington (Eds.), *Who learns what and how: The research base for teaching and learning with cases* (pp. 201–228). Mahwah, NJ: Lawrence Erlbaum Associates.

Guilar, J., & Loring, A. (2008). Dialogue and community in online learning: Lessons from Royal Roads University. *Journal of Distance Education, 22*(3), 19–40.

Gainsburg, J. (2008). Real-world connections in secondary mathematics teaching. *Journal of Mathematics Teacher Education, 11*(3), 199–219. doi:10.1007/s1085 –007-9070-8

Garii, B., & Okumu, L. (2008). Mathematics in the world: What do teachers recognize as mathematics in real-world practice. *Montana Mathematics Enthusiast, 5*(2–3), 291–304.

Glaser, B. G., & Strauss, A. L. (1967). *The discovery of grounded theory: Strategies for qualitative research.* New York, NY: Aldine.

Gutstein, E. (2006a). *Reading and writing the world with mathematics: Toward a pedagogy for social justice.* New York, NY: Routledge.

Gutstein, E. (2006b). The real world as we have seen it: Latino/a parents' voices on teaching mathematics for social justice. *Mathematical Thinking and Learning, 8*(3), 331–358.

Huang, H. E. (2004). The impact of context on children's performance in solving everyday mathematical problems with real-world settings. *Journal of Research in Childhood Education, 18*(4), 278–292. doi:10.1080/02568540409595041

Kale, U., Brush, T., & Saye, J. (2009). Assisting teachers' thinking and participation online. *Journal of Educational Computing Research, 41*(3), 287–317.

Lichtman, M. (2011). *Understanding and evaluating qualitative educational research*. Thousand Oaks, CA: Sage Publications.

Loo, S. (2006). Adult numeracy teacher training programmes in England: A suggested typology. *International Journal of Lifelong Education, 25*(5), 463–476.

Major, C. H. (2010). Do virtual professors dream of electric students? University faculty experiences with online distance education. *Teacher College Record, 112*(8), 2154–2208.

National Council of Teachers of Mathematics. (2000). *Principles and standards for school mathematics*. Reston, VA: Author.

National Governors Association Center for Best Practices, Council of Chief State School Officers. (2009). Common core standards. www.corestandards.org/FAQ. htm

Pugalee, D. K., Hartman, K. J., & Forrester, J. H. (2008). Performance of middle grades on quantitative tasks: A beginning dialogue on quantitative literacy in middle schools. *Investigations in Mathematics Learning, 1*(2), 35–51.

Smith, J. A., & Sivo, S. A. (2012). Predicting continued use of online teacher professional development and the influence of social presence and sociability. *British Journal of Educational Technology, 43*(6), 871–882.

Steen, L. A. (2001). The case for quantitative literacy. In L. A. Steen (Ed.), *Mathematics and democracy: The case for quantitative literacy* (pp. 1–22). Princeton, NJ: Woodrow Wilson Foundation.

Steenken, W. (2007). Quantitative literacy: Beyond crossroads gets it right. *AMATYC Review, 28*(2), 22–25.

Vlachopoulos, P., & Cowan, J. (2010). Reconceptualising moderation in asynchronous online discussions using grounded theory. *Distance Education, 31*(1), 23–36. doi:10.1080/01587911003724611

Verschaffel, L., Greer, B., & DeCorte, E. (2000). *Making sense of word problems*. The Netherlands: Swets and Zeitlinger.

Wiest, L. R., Higgins, H. J., & Frost, J. H. (2007). Quantitative literacy for social justice. *Equity and Excellence in Education, 40*(1), 47–55. doi:10.1080/10665680601079894

Wilder, E. I. (2010). A qualitative assessment of efforts to integrate data analysis throughout the sociology curriculum: Feedback from students, faculty, and alumni. *Teaching Sociology, 38*(3), 226–246. doi:10.1177/0092055X10370118

Wilkins, J. M. (2000). Preparing for the 21st century: The status of quantitative literacy in the United States. *School Science and Mathematics, 100*(8), 405–417.

Wills, J. B., & Atkinson, M. P. (2007). Table reading skills as quantitative literacy. *Teaching Sociology, 35*(3), 255–263.

Chapter 6

Something Doesn't Add Up

Math Teachers and Student-Centered Pedagogy

David Nurenberg and Se-Ah Kwon Siegel

INTRODUCTION

At times it seems the more research tells us about what engages students in learning, the more education policy pushes us in precisely the opposite direction. Policymakers have set as goals for all students the development of "21st-century competencies" and higher-order thinking skills. While these skills, such as critical thinking and problem solving, "have been components of human progress throughout history, what's actually new is the extent to which changes in our economy and the world mean that collective and individual success depends on having such skills" (Rotherham & Willingham, 2009, par. 2).

Unfortunately for public education, the means by which most schools attempt to address those goals—by choice, by imposition, or by lack of training—have been insufficient, even retrograde, to that mission. These means include the standardized tests used to assess, reward, and punish students, teachers, and administrators alike. What Brooks and Brooks (1999) wrote almost twenty years ago remains true today:

> Many school districts . . . are searching for broader ways for students to demonstrate their knowledge. However, the accountability component of the standards movement has caused many districts to abandon performance-based assessment practices and refocus instead on preparing students for paper-and-pencil tests. The consequences for districts and their students are too great if they don't. (par. 17)

In turn, teacher training programs are faced with an ethical dilemma: if they choose to prepare their graduates to understand, develop, and employ pedagogies that engage students and prepare them for high-level analysis and critical thinking, are they underpreparing future teachers for the kinds of accountability structures, and associated professional pressures, that will compel them to teach in a more proscribed and limited way? Once a new teacher is faced with a decision between adhering to a certain practice, however well-supported by research, and preserving his or her job and ability to remain an influence on students, we should not be surprised if they abandon the former in favor of the latter.

REVIEW OF THE LITERATURE

Since the late 1990s, there has existed "a general—if somewhat loose and shifting—consensus" (Great Schools Partnership, 2016, par. 4) that schools must teach skills that are "essential to prepare all students for the challenges of work, life, and citizenship in the 21st century and beyond" (Partnership for 21st Century Learning, 2016, par. 4). Skills such as analysis and critical thinking, problem solving, synthesizing and evaluating information, and working successfully in groups are seen necessary for "thriv[ing] in a world where change is constant and learning never stops" (par. 1).

This mission, however, runs up against mandates of the No Child Left Behind Act (NCLB; 2001) and its ideological descendants like Race to the Top with their emphasis on standardized accountability measures for students, teachers, and schools, which have pushed schools toward "teaching skills and content in the format of the test only, drilling students on specific skills and content areas that will be on the test, and spending more class time preparing for testing" (Rowland, 2011, p. 3). At its most extreme, this resulted in situations where "nobody believed test-drilling was of educative worth. Its only function was to defend the school from state or federal punishments" (J. Kozol in National Education Association [2007, p. 25]).

Research is overwhelming that such methods have a chilling effect on student engagement. Jimerson, Campos, and Greif (2003) define school engagement as the "affective, behavioral, and cognitive dimensions" of students' attitudes toward school and learning (p. 1). Yazzie-Mintz and McCormick (2012) define engagement as "a function of the perceptions of students about their experiences in the learning environment" (p. 743). Student engagement has been consistently linked with academic success at all grade levels, particularly the middle and high school years (Mulkey, Catsambis, Steelman, & Crain, 2005; Newmann, 1992; Skinner & Belmont, 1993; Reyes, Brackett, Rivers, White, & Salovey, 2012; Upadyaya & Salmela-Aro, 2013; Wang & Holcombe, 2010), even if it is not the sole determinant (Dotterer & Lowe, 2011).

Are students in the post–NCLB world engaged? The nationally administered 2009 High School Survey of Student Engagement ($N = 42,754$) revealed that 49% of students reported being bored at some point every day and 17% reported being bored every class (E. Yazzie-Mintz in Indiana University Newsroom [2010]). Among the top reasons students reported for considering dropping out was "I didn't see the value in the work I was being asked to do" (42.3%). One student respondent said, "My school is a good school, but I feel the only thing they care about are scores on standardized tests, not us students as individuals" (Yazzie-Mintz & McCormick, 2012, p. 743).

Eighty-two percent of student respondents agreed or strongly agreed that "opportunities to be creative at school" motivate them more, and 65% agreed or strongly agreed that questions without singular "right or wrong" answers were more engaging (par. 6). This trend even holds among students labeled as "gifted" (Kanevsky & Keighley, 2003).

Methods that have a better track record for student engagement include various forms of inquiry teaching and learning (Jablon, 2014), cooperative learning (Baloche, 1998; Johnson & Johnson, 2005), project-based learning (Larmer, Mergendoller, & Boss, 2015; Markham, 2003), student-centered classroom management (Burden, 2016; Lieber, 2009), and problem-based learning (Delialioğlu, 2012).

What all these methods have in common is their constructivist approach to learning (Vygotsky in Van der Veer & Valsiner [1991]), which positions students in the role of co-facilitators of their own education. Rather than expecting students to "bank" knowledge from lectures or readings, such methods aim to help students come to or create knowledge on their own, supported by and developed through interactions with materials, classmates, and teachers.

Collectively, these methods can often be grouped under the heading Student-Centered Approaches (SCAs). This term refers to pedagogies and structures that "mov[e] students from passive receivers of information to active participants in their own discovery process" (International Society for Technology in Education, 2017, p. 1).

Student-centered learning is "at its core . . . about finding ways of teaching that truly engages a student in education" (Virtual Learning Academy, 2017, p. 1). These are not obscure or arcane ideas nor are they particularly novel:

> These approaches are widely acclaimed and can be found in any pedagogical methods textbook; teachers know about them and believe they're effective. And yet, teachers don't use them. Recent data show that most instructional time is composed of seatwork and whole-class instruction led by the teacher. (Rotherham & Willingham, 2009, par. 20)

Rotherham and Willingham blame this phenomenon largely on inadequate teacher training. But if teacher preparation were the only issue, then a school such as Lesley University, which focuses its teacher education programs on such methods explicitly, should produce graduates who go on to use them in the field. Yet our study revealed that this is not always so.

It is possible that even teachers confident in using student-centered methodologies may decide not to use them. One reason may be that while engagement and achievement are related, research on student-centered teaching's direct, value-added impact on academic performance remains mixed. Some meta-analyses of the research support positive impact on achievement (e.g., Baeten, Dochy, & Struyven, 2013; Bonwell & Eison, 1991; Freeman et al., 2014; Prince, 2004), but others find superiority in traditional methods (e.g., Chall, 2000). Prince (2004) observes that "confusion can result from reading the literature on the effectiveness of any instructional method unless the reader and author take care to specify precisely what is being examined" and even aggregated meta-analyses of the research "can be misleading if the forms of [pedagogy] vary significantly in each of the individual studies included in the meta-analysis," which is inevitably the case (pp. 1–2).

While it was beyond the scope of this study to produce a definitive determination of the impact of student-centered pedagogies on student achievement, we wish to call attention to the "significant problem with assessment" that Prince (2004) describes in that "many relevant learning outcomes are simply difficult to measure. This is particularly true for some of the higher level learning outcomes that are targeted by active learning methods" (p. 2).

As Brooks & Brooks (1999) put it,

the efficacy of the learning environment is a function of many complex factors, including curriculum, instructional methodology, student motivation, and student developmental readiness . . . Trying to capture this complexity on paper-and-pencil assessments severely limits knowledge and expression . . . [and] contravenes what years of painstaking research tells us about student learning . . . Classroom practices designed to prepare students for tests clearly do not foster deep learning that students apply to new situations. Instead, these practices train students to mimic learning on tests. (par. 13–15)

If the easiest way to measure student achievement only measures a narrow band of skills, especially skills that are not 21st-century competencies, and if schools in the post–NCLB era are evaluated mainly via these limited methods (Owens & Sunderman, 2006), then there may be a perverse incentive for even highly capable teachers to eschew the student-centered, active-learning strategies most likely to teach students 21st-century skills (Au, 2007; Valli & Buese, 2007). Anecdotes from Lesley College's recent graduates—graduates claiming direct experience in that very phenomenon—led us to undertake this study.

GOALS OF THE STUDY

Since its inception as Lesley College in 1909, Lesley University's education schools and associated teacher training programs have been "deeply committed to student-centered and client/person-centered pedagogy" and have placed great emphasis on training teacher candidates in these methods (Lesley University NEASC Self-Study, 2015, p. 81). As a core faculty member and an associate director of assessment, the two authors of this study sought to determine who among Lesley's graduates have indeed gone on to employ (or not to employ) the student-centered pedagogies they were trained in; to what extent; and for what reasons they either use or do not use these pedagogies in their teaching practice.

METHODOLOGY

We constructed and conducted a 27-item survey that we sent out via email to 34,757 individuals who had graduated from Lesley University's programs from 1970 to 2015 and who had email addresses in the alumni database. We received 329 responses, and 86 initial respondents exited the survey, as instructed, because they were not currently active classroom teachers in the United States. This study focuses on the remaining respondents ($n = 243$).

1. Demographic data from these graduates
2. Their self-reported data regarding
   a. what methods of teaching they employed
   b. how effective they judged these methods to be
   c. what assessment tool or tools they used to made this judgment

The survey presented 25 separate teaching methods or practices using terminology taken from various syllabi in Lesley's teacher preparation courses. We then divided those methods into two categories. We titled one category, composed of 20 methods, "Student-Centered Approaches" (SCAs; see Table 6.1 for the full list of these items). We ran a reliability analysis of those 20 items and found a Cronbach's alpha of 0.894 for them. We calculated a usage score by averaging the scores of all 20 of the SCA items. The higher the score, the more frequently the teachers used student-centered approaches.

Table 6.1. Frequency of Reported Use of SCAs by All Respondents

#	Question	Never	Rarely	Sometimes	Often	All the time	I don't know/I am not sure	I used to but don't anymore	Total
1	Inquiry level 2 learning—Students develop their own approaches to solve a problem given to them.	8	19	86	112	28	5	1	259
2	Inquiry level 3 learning—Students develop their own problems as well as their own approaches to solve the problem.	27	49	97	66	13	3	2	257
3	Project-based learning—Students learn new skills and knowledge through pursuing a project (Note: this is different from "doing a project" at the end of a unit as a summative assessment)	20	41	68	75	51	4	2	261
4	Working in cooperative groups	3	7	53	114	83	1	0	261
5	Class discussions, full class	4	13	52	115	75	0	2	261
6	Socratic circles	62	49	65	43	11	25	0	255
7	Think-pair-share exercises	25	30	52	99	52	0	0	258
8	Class meetings	42	42	76	52	34	11	0	257
9	Negotiating classroom rules and protocols with the students	30	42	67	69	42	2	1	253
10	Students construct their own lessons and teach classmates or younger students	62	73	76	31	8	2	2	254
11	Students peer-edit one another's work	36	38	93	74	15	1	0	257
12	Portfolio-based assessment	63	42	66	57	25	2	0	255

#	Statement								
13	Assignments that engage students in self-reflection and self-evaluation of their own work	12	24	89	97	34	0	1	257
14	Students play a role in creating teaching resources	55	71	85	37	3	1	0	252
15	Assignments that apply learning to real-world situations	6	18	67	106	58	0	0	255
16	Service learning—Classroom work involves, in a meaningful way, interaction with the larger community (school, town, etc.)	56	68	78	38	13	2	0	255
17	"Flipped classroom"—In-class time is used for workshop while lectures/PowerPoints/videos are the homework	132	55	38	24	2	2	1	254
18	Student interests and personal, out-of-classroom life experiences are routinely referenced and connected to classroom lessons	5	17	77	104	52	0	0	255
19	The teacher serves as "guide on the side" more than "sage on the stage"	8	21	84	97	45	2	0	257
20	Peer mediation/negotiation/conflict resolution	45	50	77	60	19	4	0	255

The other category we created was labeled "Traditional Methods." Our analysis will use "Lecture/PowerPoint" to represent traditional methods for our study as this is the only approach that does not require the active participation of students, distinguishing it from the main trait of student-centered approaches.

While the survey was designed to be anonymous, respondents had the option of leaving contact information for follow-up interviews. A total of 170 respondents chose to provide such information. Because of the significant results our analysis obtained regarding math teachers, we attempted follow-up interviews with the 14 math teachers who left their contact information, and we conducted interviews, via phone and email, with four of them.

ANALYSIS OF THE DATA

The principal investigators for this study conducted an analysis of data following a five-step process. The following sections present a discussion of the analysis process aligned with data tables demonstrating the data analysis.

Part I: Descriptive Statistics

Of the 243 alumni examined, 68% were employed in Massachusetts and the rest elsewhere in the United States, and 80% completed their programs between 2000 and 2015. We conducted independent sample T and ANOVA tests to determine significant differences in frequency of use of student-centered methods among teacher groups. We ran a multiple linear regression to identify teacher groups that had a statistically significant contribution in predicting the frequency of use of student-centered methods. Table 6.2 summarizes the characteristics of the sample group and descriptive statistics.

Part II: Use of Lecture/PowerPoint vs. SCAs

Of the sample, 36.2% of the teachers reported that they never used lecture/PowerPoint or rarely used lecture/PowerPoint while 36.6% reported using it often or all of the time (see Table 6.3).

The largest demographic by far of teachers who reported "never or rarely" using SCAs (50%) was among those who had five or fewer years of teaching experience (see Table 6.4). Approximately 25.8% of math teachers reported "never or rarely" using lectures (see Table 6.5) in comparison with 54% of pre-K and elementary teachers who reported lecturing "never or rarely" (see Table 6.6).

Table 6.2. Characteristics of the Survey Population

Age Distribution		
Age	Frequency	%
20–30	41	16.90
30–40	64	26.30
40–50	69	28.40
50+	68	28.00
Missing	1	0.40
Total	243	100.00

Gender Distribution		
Gender	Frequency	%
Male	40	16.50
Female	202	83.10
Missing	1	0.40
Total	243	100.00

Distribution of Teachers by Years of Teaching Experience		
Teaching Experience	Frequency	%
5 Years or less in Teaching	64	26.30
5–10 Years of Experience	41	16.90
10–20 Years of Experience	84	34.60
20+ Years of Experience	54	22.20
Total	243	100.00

Distribution of Teachers by Teaching Subject		
Subject	Frequency	%
All subjects (Pre-K/K/ Elementary)	85	35.00
ELA or Humanities	41	16.90
Math	31	12.80
Science	13	5.30
Foreign Language	6	2.50
Other	67	27.60
Total	243	100.00

Distribution of Teachers by Grade Level Taught		
Grade Level Taught	Frequency	%
Pre-K–K	43	14.78
Elementary	106	36.43
Middle School	54	18.56
High School	58	19.93
Post-Secondary	30	10.31
Total*	291	100.00

* Teachers can pick more than one grade level if that applies to them

Table 6.3. Frequency of Reported Use of Traditional Methods by All Respondents

Teaching Method Response Category	Lecture/PowerPoint	
	N	%
Never	41	16.9
Rarely	47	19.3
Sometimes	63	25.9
Often	59	24.3
All the time	30	12.3
I don't know/I am not sure	0	0
I used to but don't anymore	2	0.8
Subtotal	242	99.6
Missing	1	0.4
Total	243	100

Table 6.4. Distribution of Responses to How Often Teachers Reported Using SCAs, by Years of Teaching Experience

N of Teaching Experience	%	Rarely or Never	Sometimes	Often or All the time	Total
5 or less	Count	32	30	2	64
	%	50.0%	46.9%	3.1%	100%
5–10	Count	14	24	2	40
	%	35.0%	60.0%	5.0%	100%
10–20	Count	30	44	10	84
	%	35.7%	52.4%	11.9%	100%
20+	Count	19	29	6	54
	%	35.2%	53.7%	11.1%	100%
All	Count	95	127	20	242
	%	39.3%	52.5%	8.3%	100%

Table 6.5. Self-Reported Use of Lecture/PowerPoint among Math Teachers

Level of Use	Frequency	%	Cumulative %
Never	3	9.7	9.7
Rarely	5	16.1	25.8
Sometimes	9	29.0	54.8
Often	10	32.3	87.1
All the time	3	9.7	96.8
I used to but don't anymore	1	3.2	100.0
Total	31	100.0	

Table 6.6. Self-Reported Use of Lecture/PowerPoint among Pre-K and Elementary Teachers

Level of Use	Frequency	%	Cumulative %
Never	25	29.41	29.41
Rarely	20	23.53	52.94
Sometimes	13	15.29	68.24
Often	16	18.82	87.06
All the time	9	10.59	97.65
I used to but don't anymore	1	1.18	
Subotal	84	98.82	98.82
Missing	1	1.18	100.00
Total	85	100.00	

We found that 25.8% of math teachers reported never or rarely using lecture/PowerPoint (see Table 6.5) whereas 52.9% of pre-K/elementary teachers responded this way (see Table 6.6). In addition, 43% of the math teachers said that they lectured "often" or "all the time" (see Table 6.5) compared to 29.4% of the pre-K/elementary teachers who reported this (see Table 6.6).

Table 6.7 shows that 25.9% of pre-K and elementary teachers reported using SCAs "never or rarely" whereas 64.5% of math teachers reported using them "never or rarely."

Table 6.7. Self-Reported Use of SCAs: Pre-K, Elementary School Teachers, Math Teachers

Teacher Group		Level of SCA Use			
		Rarely or Never	Sometimes	Often or All the time	Total
Pre-K/K/Elementary	Count	22	53	10	85
	% within Group	25.9%	62.4%	11.8%	100%
Math	Count	20	9	2	31
	% within Group	64.5%	29.0%	6.5%	100%

To determine the statistical difference in percentages in the responses, one-way ANOVA testing was performed to look at comparisons by teaching subject area, finding statistically significant differences between math teachers and pre-K/elementary teachers. As shown in Table 6.8, on average, pre-K/elementary teachers used SCAs more often than math teachers ($p = 0.037$).

Table 6.8.   ANOVA Test of SCA Scores

	Sum of Squares	Df	Mean Square	F	Sig.
Between Groups	5.476	5	1.095	2.969	.013
Within Groups	87.038	236	.369		
Total	92.514	241			

Table 6.9 shows subject group comparisons between elementary teachers teaching only math versus elementary teachers teaching multiple subjects. Again, elementary teachers teaching multiple subjects used SCAs more often than elementary teachers teaching only math.

Table 6.9.   Subject Group Comparisons

(I) What is the subject you are currently teaching? (If you are teaching multiple subjects, pick the . . .)	(J) What is the subject you are currently teaching? (If you are teaching multiple subjects, pick the . . .)	Mean Difference (I–J)	Std. Error	Sig.	95% Confidence Interval Lower Bound	Upper Bound
Math	All subjects (Pre-K/K/ Elementary)	–.44319*	.12742	.037	–.8708	–.0156
	ELA or Humanities	–.29334	.14454	.534	–.7784	.1917
	Science	–.16367	.20067	.985	–.8371	.5097
	Foreign Language	–.64997	.27086	.334	–1.5589	.2590
	Other	–.30004	.13223	.401	–.7438	.1437

*The mean difference is significant at the 0.05 level.

As shown in Table 6.10, even when compared to other secondary school teachers, math teachers at the secondary level still used SCAs less often than other teachers of other subjects at the secondary school level ($p = 0.003$).

Table 6.10.   Descriptive Statistics: Secondary Math Teachers vs. Secondary Teachers of Other Subjects

	Teacher Group	N	Mean	Std. Deviation	Std. Error Mean
SCA Score	Secondary Math Teachers	24	2.7140	.67792	.13838
	Secondary Teachers of Other Subjects	83	3.1885	.65923	.07236

Since a general belief exists that more "playful" (open-ended, student-directed) educational models are "limited to just primary and early years education" while "the 'fun' end[s] when we graduate to grown-up school" and "creative and imaginative teaching methods" are dispensed with (Jenkin, 2013, par. 10), we ran an additional, independent *T*-test to see if, when compared with the other secondary school educators in our survey, math teachers still demonstrated statistically significant rarity in using SCAs (see Table 6.11). We found that, yes, even when compared to other secondary school teachers, math teachers at the secondary level still used SCAs less often than other teachers of other subjects at the secondary school level ($p = 0.003$).

Table 6.11.   Independent Sample T-Test, Difference: Secondary Math Teachers vs. Secondary Teachers of Other Subjects

	T	Df	Sig. (2-tailed)	Mean Difference	Std. Error Difference	95% Confidence Interval of the Difference	
						Lower	Upper
SCA Score	−3.086	105	.003	−.47451	.15375	−.77935	−.16966

In Table 6.12, the regression results show the positive relationships between teachers of mathematics and SCA use and teaching experience and SCA use. The explanatory power of the model is small, with adjusted R Square = 0.057 with $p = 0.001$. However, both teaching experience and teachers of mathematics are statistically significant predictors for the use of SCAs.

Table 6.12. Significance of Predictors by Regression Results

		Linear Regression Table				
Dependent Variable: SCA Scores, Frequency Scale Average in a Combination of SCAs		*Unstandardized Coefficients*		*Standardized Coefficients*		
		B	*Std. Error*	*Beta*	*T*	*Sig.*
Predictors	Teachers of mathematics	0.342**	0.125	0.184	2.73	0.007
	Elementary/ Secondary teachers	−0.041	0.084	-0.033	−0.492	0.623
	Number of years teaching experience	0.012**	0.005	0.179	2.734	0.007

**p value is less than 0.01

The contributing value for teachers of mathematics is found in a standard coefficients beta of 0.184 with a standard beta coefficient of 0.179 for teaching experience. Both of the independent variables have a similar level of influence on the outcome.

In summary, teachers of mathematics were likely to use student-centered methods less frequently than teachers of other subjects, and novice teachers were likely to use student-centered methods less frequently than experienced teachers did.

Part III: Self-Reported Reasons for Using SCAs vs. Traditional Methods

When participants were asked, "If you use any of those student-centered methods/techniques/structures, why do you use them?" Table 6.13 reflects those who responded, "They engage students more," and Table 6.14 reflects those who responded, "They increase student learning and achievement."

Table 6.13. Distribution of Teacher Self-Reports on SCA Effect on Student Engagement

Teacher Group Separated by SCA Usage Mean Score	*N*	*%*
Teacher Group 1 (Never–Rarely)	95	86.3
Teacher Group 2 (Sometimes)	127	95.3
Teacher Group 3 (Often–All the time)	20	100

Table 6.14. Distribution of Teacher Self-Reports on SCA Effect on Student Learning and Achievement

Teacher Group Separated by SCA Usage Mean Score	N	%
Teacher Group 1 (Never–Rarely)	95	74.7
Teacher Group 2 (Sometimes)	127	92.9
Teacher Group 3 (Often–All the time)	20	100

Table 6.15 reflects by what metric those respondents reporting SCAs increased student learning and achievement determined that increase. In terms of frequency, in-class assessments (both informal observations and formal work) as well as student self-reports constituted the primary means of determining that SCAs had a positive impact on student achievement. Only a very few respondents reported that standardized tests were among the means by which they saw this impact on achievement. Of the teachers who chose increased student achievement as a result of SCAs, no statistically significant differences were found at any grade or subject level.

Table 6.15. Assessment Types Chosen by Teachers Who Reported That SCAs Increased Student Achievement

Assessment Type/ Teacher Group	Teacher Group 1 (Never–Rarely)	Teacher Group 2 (Sometimes)	Teacher Group 3 (Often–All the time)
In-class assessments (tests, quizzes, essays, problem sets, etc.)	55 (57.9%)	97 (76.4%)	13 (65.0%)
MCAS scores	4 (4.2%)	7 (5.5%)	NA
College-related standardized tests (AP exams, Achievements, SATs, etc.)	3 (3.2%)	7 (5.5%)	1 (5.0%)
Other types of standardized tests	8 (8.4%)	31 (24.4%)	4 (20.0%)
Student self-reports	39 (41.1%)	78 (61.4%)	13 (65.0%)
In-class observations	63 (66.3%)	116 (91.3%)	18 (90.0%)

Table 6.16 summarizes the reasons that math teachers reported for electing not to use SCAs. In terms of frequency, difficulty of design or execution was the most-selected response by nearly a factor of three. Several of the "Other/Please explain" responses involved a reference to the approaches as "not being appropriate" for their particular students due to either academic or behavioral issues.

Table 6.16.   Why Did You Stop Using SCAs? (Math Teachers Specifically)

Reason	Frequency	%
Students were not engaged.	3	13.04%
Students were not learning or achieving well.	3	13.04%
They were too difficult to design or employ.	7	30.43%
An administrator or team leader asked me to stop.	2	8.70%
Other/Please explain:	8	34.78%
Total	23	

Part IV: Qualitative Interview Follow-up with Math Teacher Respondents

Fourteen math teacher respondents who reported minimal use of SCAs listed their information for follow-up, and we contacted them all via email and phone. We were able to conduct interviews with four of them. In three of the four interviews, respondents reported pressure for "pace of coverage," either directly from administrators or self-imposed, as a major cause. For example, while one respondent reported that she personally found "inquiry-based learning . . . really does engage students in their learning in a way that other approaches do not," she added that "with the sheer amount of standards that we need to cover and the shifting curriculums . . . I have not been able to fit in as much as I would like."

Another interviewee self-identified as a teacher "who must teach so very much in a school year . . . I find that these [student-centered] methods honestly take too long." The third respondent explained that "the board approved textbooks that [administrators] said we had to use . . . don't lend well to inquiry based teaching . . . The textbook just has them follow step by step instructions." That same respondent summed up her dilemma as follows: "I have to find in my own way a happy medium between what is best for the students and what will not get me fired."

Administrative pressure was not the only factor the interviewees mentioned. Two of the respondents mentioned that they felt their students' learning needs were too great to accommodate SCAs: "It's pretty hard to discover something," one respondent said, referring to inquiry and discovery learning, "if you don't have the foundation for what you're supposed to be discovering."

Part V: Lesley University Teacher Preparation for Using SCAs

As illustrated in Table 6.17, 84.6% of respondents reported that their teacher training program at Lesley focused on SCAs at least some of the time, with

45.8% reporting the focus as being "Often" or "All the time." We found no significant difference between secondary math teachers and secondary teachers of other subjects or between elementary teachers and secondary teachers in terms of reporting that their program covered these methods.

Table 6.17.   How Often Did Your Program at Lesley Include Training/Focus on SCAs?

Level of Usage	Frequency	%
Never	7	8.2
Rarely	5	5.9
Sometimes	33	38.8
Often	28	32.9
All of the time	11	12.9
I don't know/I'm not sure	1	1.2
Total	85	100.0

DISCUSSION

More than 80% of respondents reported that their teacher training program at Lesley spent a great deal of time on SCAs. This increases the confidence with which we can say that the respondents to our survey were at the very least exposed to the skills and key competencies needed for understanding, designing, and employing SCAs.

The statistically significant correlation between teaching in the early grades and using SCAs would seem to support the "accepted knowledge" of the greater use of student-centered techniques with younger students versus older ones (Brown et al., 2004; Lamb-Sinclair, 2016). The use of SCAs seems to be rarer after the elementary grades despite active efforts to change this as "the change challenge is profoundly different in elementary vs. secondary education" (Newell, 2003, p. vii).

This may be attributable to high school's stronger historical ties to the factory-line production model of education (Spring, 2017) or to pressure being greater to do well on high-stakes tests in middle and high school or even to a general sense that "serious" teaching and learning, of the kind that distinguishes the higher grades from the primary ones, is more teacher directed (Jenkin, 2013).

Whatever the cause, this disparity is tragically ironic as there is evidence that SCAs have an even greater positive effect on student achievement in secondary schools versus elementary schools (Gage, 1978). What is undeniably clear in the literature, including a 2015 Gallup sampling of over

900,000 students, is that student "engagement levels . . . show a consistent decrease as students get older, bottoming out in 11th grade" (Gallup [2015] in Brenneman [2016, p. 1]). *Something* has changed for students in the higher grades in regard to their engagement with their schooling experience, and the shift away from student-centered pedagogy may be a factor.

Our finding that teachers with less experience use SCAs less frequently than their more veteran colleagues both confirmed and challenged our predictions. On the one hand, we had expected that veteran teachers would have a greater repertoire of pedagogical and management techniques and be more confident in employing them. But we had also expected that younger or more recent graduates[1] might be less set in their ways and less wedded to "traditional" methods like lecture and drills having grown up knowing the "21st-century" world. However, these are also the same teachers who were themselves a product of NCLB–influenced education and therefore their own schooling experience may have been more rote and standardized-test-focused than that of their predecessors.

"Lack of effective models" is an obstacle for all teachers attempting student-centered teaching, but this could be particularly dissuasive for newer teachers (Jablon, 2014, p. 11). Newer teachers are also presumably more influenced by administrative pressure. A nationwide National Education Association survey of 1,500 teachers yielded 72% replying that they felt "moderate to extreme" pressure from school and district administrators and that 40% reported their school placing "moderate to extreme" emphasis on student test scores to evaluate their performance (Walker, 2014, par. 3). Although policies differ from state to state, most teachers nationally do not gain the protection of tenure until their fourth year, so it is the novices who may be less willing to take risks with unfamiliar methodologies (National Council on Teacher Quality, 2015).

But our study found that even among secondary school teachers of all experience levels, math teachers seem particularly indisposed to use SCAs. Since respondents could check multiple reasons for choosing not to use SCAs, the responses in Table 6.16 may all be sides of the same coin. For our math teacher respondents, SCAs may have not engaged their students because they were too difficult to employ correctly, or perhaps it was the reverse and SCAs were too difficult to employ because students were not engaged and would not stay on task. Either way, student achievement seemed to have suffered, in the respondents' view.

The interviewees' reports that student learning needs *precluded* the use of SCAs may seem puzzling to those who are persuaded by the research linking such methods to increased student engagement and achievement. Wouldn't students experiencing the most difficulty in mathematics be precisely the population to use SCAs with? While the "math wars" between so-called traditional and progressive approaches to teaching mathematics have been going on since the 1990s (Schoenfeld, 2004), similar debates continue to take place across all academic disciplines (Cothran, 2016). What is special about math that it could dissuade even a teacher trained in SCAs to elect not to use them?

Part of the reason may stem from the perception that SCAs provide "students with the analytical tools they need to succeed on English/language arts standardized tests . . . [But] both teaching mathematics in general, and teaching it so students succeed on state and national benchmarks, is harder to do in an inquiry-driven fashion" (Quillen, 2013, par. 6). Mathematics educators have historically viewed SCAs as "overly permissive" and lacking in "rigor" (Brooks & Brooks, 1999, par. 33–35) or even as an "intellectual anarchy that lets students pursue whatever interests them and invent and use any mathematical methods they wish, whether those methods are correct or not" (Battista, 1999, p. 429).

Yet a body of research supports the benefits of successfully using SCAs specifically for mathematics instruction (Fennema et al., 1996; Freeman et al., 2014; Gonzales et al., 2004; Panitz, 2000; Polly, 2008; Nurenberg, 2010; Polly et al., 2013; Polly et al., 2015; Wang, Brinkworth, & Eccles, 2013). One of the larger studies ($N = 688$) found that "students whose teachers had reported teacher-centered beliefs and teacher-centered practices had significantly lower gain scores on curriculum-based assessments" (Polly et al., 2015, p. 23). Even in some studies where mathematics achievement showed no particular improvement when taught via student-centered interventions, students at least reported greater engagement and a more positive attitude (Clark, 2015; Gningue, Peach, & Schroder, 2013).

Of course, the above research is subject to Prince's (2004) cautions: each of the studies cited here dealt with a different subset of SCA skills (cooperative learning, inquiry, and flipped classroom). That said, the fact that so many of these sub-skills seem able to produce improved results may suggest that no single one of them is the "essential" best practice. Indeed, some research indicates that SCAs tend to work best when used in concert (Jablon, 2014).

Yet another factor governing math teachers' decision whether or not to employ SCAs may be the standing recommendations of the National Mathematics Advisory Panel—whose members included Harvard University, the Carnegie Foundation for the Advancement of Teaching, and the U.S. Department of Education—that advised that "all-encompassing recommendations that instruction should be entirely 'student centered' or 'teacher directed' are not supported by research" (National Mathematics Advisory Panel, 2008, p. 45). Boaler (2008) and others criticized the panel's report as under-examining student-centered pedagogies due to a narrow and inaccurate definition as practices where "students are primarily doing the teaching" (NMAP, 2008, p. 45), which Boaler (2008) felt was a "trivialized definition" that was "inadequate" (p. 589). Furthermore, the panel only chose eight studies upon which to base their recommendation, some of which had only "monitored a particular teaching approach used for a few days" (Boaler, 2008, p. 590). Boaler argues that had the panel conducted a truly comprehensive review and analysis of research on SCAs, they might have found a more persuasive case for their use.

Perhaps the most compelling reason mathematics teachers might have for continuing to use traditional methods is that according to the dominant metrics, they are being successful. Data from the 2015 Trends in International Mathematics and Science Study revealed that

> U.S. fourth-graders' average mathematics scores increased overall from 1995 and 2015, going from 518 to 539 points. Meanwhile average math scores for U.S. eighth-graders increased from 492 points in 1995 to 518 points in 2015. (Stephens et al., 2016, in EdSurge, 2016, p. 1)

Even the more modest National Assessment of Educational Progress (NAEP) mathematics assessments reported that "average mathematics scale scores for students in grades four and eight improved almost every year between 2000 and 2013" and that twelfth graders' scores had not fallen since 2009 (NAEP in Child Trends, 2015, p. 2).

There is also recent evidence from a massive ($N = 653,316$) study that suggests that teacher impact on math scores, while more immediate than that of English language arts teachers, has less long-term impact on, and less generalizability to, achievement in other subjects (Master, Loeb, & Wyckoff, 2017). This may reinforce the idea that mathematical knowledge requires a distinct skill set not subject to the same teaching conditions as other disciplines. However, it may also, as the study's authors suggest, reward math teachers, administrators, and policymakers for privileging strategies for short-term test success rather than strategies for teaching more long-term, comprehensive knowledge.

If the traditional methods most math teachers are using seem to be working in that they achieve the results needed on the necessary schedule, where is the incentive to change? Such incentives may indeed exist. Despite gains in domestic standardized test scores, American students do not perform comparatively as well on the Program for International Student Assessment (PISA), a tri-annual international assessment of over 500,000 students from 72 countries (Ouyang, 2016). "The most recent PISA results, from 2015, placed the U.S. an unimpressive 38th out of 71 countries in math and 24th in science" (Desilver, 2017, p. 1).

The PISA exam challenges students to apply mathematical concepts to real-world situations as opposed to abstract, isolated problem sets. Students must be able to "formulate situations mathematically . . . employ mathematical concepts, facts, procedures and reasoning . . . [and] interpret, apply and evaluate mathematical outcomes" (Echazarra, Salinas, Méndez, Denis, & Rech, 2016, p. 13).

By contrast, in the analysis of Stanford University education professor Linda Darling-Hammond, "the tests [Americans] use, particularly the state standardized tests, are extremely narrow. Evidence shows that they measure almost exclusively low-level skills" (Mulholland, 2015, par. 20). A 2012 Rand Corporation study found that "only three to ten percent of elementary and middle school students in the United States were administered tests that assessed deeper learning skills" (Yuan & Le [2012] in Mulholland [2015, par. 20]).

The Organization for Economic Cooperation and Development's analysis of student and teacher self-reports from the 2012 PISA tests on the kind of instruction students were receiving in their schools found that "cognitive-activation instruction has the greatest positive association with students' mean mathematics score, on average across OECD countries" (Echazarra, Salinas, Méndez, Denis, & Rech, 2016, p. 56). The OECD defined "cognitive-activation instruction" as practices represented by the following responses:

- My role as a teacher is to facilitate students' own inquiry.
- Students learn best by finding solutions to problems on their own.
- Students should be allowed to think of solutions to practical problems themselves before the teacher shows them how they are solved.
- Thinking and reasoning processes are more important than specific curriculum content. (Echazarra et al., 2016, p. 36)

Students who learned mathematics in schools that reportedly practiced such strategies showed dramatically higher mathematics performance (as much as 19 score points) on the PISA tests before and after accounting for other teaching strategies, and there was "a positive association between cognitive-activation instruction and mean mathematics performance in every country and economy that participated in PISA 2012, except Albania" (Echazarra et al., 2016, p. 56). Furthermore, "in most participating countries, teacher-directed strategies are associated with poorer student performance in the PISA assessments" (p. 65).

The OECD report is not a blanket endorsement of all SCAs—in fact, students reporting that their mathematics education mainly included working in groups, or mainly included assignments differentiated by perceived readiness level, scored "at the lower levels of mathematics proficiency" (Echazarra et al., p. 55),[2] although this factor was mitigated if some of the other cognitive-activation strategies had also been a part of their education.

Given the substantial evidence that many SCAs, or combinations of SCAs, promote these kinds of cognitive-activation strategies, American mathematics teachers may find themselves in a dilemma: the very strategies that prepare students well for the standardized tests of lower-order mathematical skills upon which graduation and teacher jobs depend may be counter-indicated for learning mathematics at deep and complex levels.

CAUTIONS

All the data we collected derived from self-reports, which depend on the respondents' accurate assessment of their teaching methodologies and their impact on students. Koziol and Burns (1986) in a survey of the research on teacher self-reports in particular found that "teachers can be accurate reporters about their instructional practice" although that accuracy was greatest when the reports centered on specific procedures or specific time periods (see also Korb [2011]). We also did not have access to the Lesley transcript data of our respondents or to their professional evaluations. That they all successfully graduated from Lesley's program should indicate that they have the knowledge to successfully employ SCAs, but we had no means beyond their diploma to assess, evaluate, or compare their skill at doing so. We recognize the caution of Rotherham and Willingham (2009) that

> there is a widespread belief that teachers already know how to [use SCAs] if only we could unleash them from today's stifling standards and accountability metrics. This notion romanticizes student-centered methods, underestimates the challenge of implementing such methods, and ignores the lack of capacity in the field today. (par. 23)

A large-scale, pseudo-experimental study with access to data such as student grades, teacher evaluations, and standardized tests is the next step required in exploring the questions and implications raised here.

CONCLUSIONS

All American public school teachers, but perhaps especially mathematics teachers, face a conundrum: how to help students both acquire 21st-century skills and achieve competency on key standardized tests. That even graduates of a university historically branded as the champion of student-centered methodologies report hesitation to actually use these methods in their classrooms is revelatory. Do the ethical responsibilities of a teacher certification program lie in promoting what is best for student learning or in helping their graduates actually keep the jobs for which they have spent tens of thousands of dollars being trained?

Of course, whether to use teacher-centered or student-centered pedagogies is not necessarily an either-or proposition. As our respondents indicated, students with poor basic math skills may not be able to easily participate in student-centered, higher-order mathematical learning. However, the boost to engagement that SCAs confer can be a vital component of gaining buy-in from reluctant learners. Synthesizing a wide variety of instructional practices is the hallmark of an effective teacher of any subject and at any grade level.

Mathematics teachers may understandably be hesitant to risk good test performance by using strategies that are difficult to employ and may not translate into success on state standardized assessments. Sooner or later, even progressive universities like Lesley may become reluctant to risk their enrollment figures by advertising themselves as places to learn those unpopular strategies.

The passage of the Every Student Succeeds Act (ESSA) in 2015 opened up more latitude for states to shape their own assessment mechanisms. Some states, like New Hampshire, are attempting to couple traditional standardized tests with the kind of "performance assessments [that] are essential to measuring higher order skills" (Darling-Hammond, 2017, p. 4). New Hampshire is piloting a program that combines locally designed portfolio assessments with regional quality control and audits, relying on expert assessors following rubrics along with mechanisms for inter-rater reliability.

Recognizing that in our present cart-before-the-horse educational climate, assessment drives instruction, New Hampshire is undertaking this experiment because its policymakers believe that such tests can "drive improvements in teaching and learning" as they "promote the use of authentic, inquiry-based instruction, complex thinking, and application of learning . . . [and] incentivize the type of instruction and assessment that support student learning of rich knowledge and skills" (Darling-Hammond, 2017, p. 40).

A system like this is far more costly to create and manage than simply contracting with a company like Pearson or ETS to produce, administer, and grade a one-size-fits-all test. It will be difficult to overcome the economic inertia of sunk costs in present systems or the ideological inertia that often forgives the questionable validity of present standardized tests because of the lure of simple, easily compared numbers.

The pressing task for future research is to expand on and illuminate whether, where, and how student-centered pedagogies impact current standardized assessments. If extensive research cannot in the end support the benefit of student-centered pedagogies for performance on these tests but simultaneously reports benefits for higher-order thinking as measured through more complex tools, then not only teachers and teacher certification programs but the American citizenry as a whole will face an enormous decision: Will we remain wedded to the current architecture of state school and student evaluation even if it means we must abandon the goal of helping students learn 21st-century skills?

Or will we adjust our metrics, perhaps to something closer to the PISA model, or to an even more versatile and comprehensive (yet potentially less generalizable) means of measuring higher-order learning like New Hampshire's? It is inescapably clear, though, that to ask our schools, our teachers, our teacher preparation programs, and most of all our students to simultaneously achieve contradictory goals is both unjust and, in the end, unsustainable.

NOTES

1. While we realize that years of teaching experience do not necessarily equate with teacher age or how recently a student graduated from Lesley's program, in most cases it does.

2. In an unfortunately confusing use of terminology, the OECD categorizes these low-scoring strategies, and only these strategies, as "student-oriented" teaching and learning. Under this article's definition of SCAs, most if not all "cognitive-activation" strategies would therefore qualify.

REFERENCES

Au, W. (2007). High-stakes testing and curricular control: A qualitative metasynthesis. *Educational Researcher, 36*(5), 258–267.

Baeten, M., Dochy, F., & Struyven, K. (2013). The effects of different learning environments on students' motivation for learning and their achievement. *British Journal of Educational Psychology, 83*(3), 484–501.

Baloche, L. (1998). *The cooperative classroom*. Upper Saddle River, NJ: Prentice Hall.

Battista, M. T. (1999). Fifth graders' enumeration of cubes in 3D arrays: Conceptual progress in an inquiry-based classroom. *Journal for Research in Mathematics Education, 30*(4), 417–448.

Boaler, J. (2008). When politics took the place of inquiry: A response to the National Mathematics Advisory Panel's review of instructional practices. *Educational Researcher, 37*(9), 588–594.

Bonwell, C. C., & Eison, J. A. (1991). *Active Learning: Creating Excitement in the Classroom. 1991 ASHE-ERIC Higher Education Reports.* Washington, DC: ERIC Clearinghouse on Higher Education, George Washington University.

Brenneman, R. (2016). Gallup student poll finds engagement in school dropping by grade level. *Education Week, 35*, 25.

Brooks, M., & Brooks, J. (1999). The courage to be constructivist. *Constructivist Classroom, 57*(3), 18–24. Retrieved from www.ascd.org/publications/educational -leadership/nov99/vol57/num03/The-Courage-to-Be-Constructivist.aspx

Brown, G., Cadman, K., Cain, D., Clark-Jeavons, A., Fenten, R., Foster, A., Jones, K., Oldknow, A., Taylor, R., & Wright, D. (2004). *ICT and Mathematics: A guide to learning and teaching mathematics.* Leicester, UK: Mathematical Association. Retrieved from https://pdfs.semanticscholar.org/a48c/2e0449a2808b5fbc39038a4 6c2ae7e2decf8.pdf

Burden, P. (2016). *Classroom management: Creating a successful K–12 learning community.* Hoboken, NJ: John Wiley & Sons.

Chall, J. S. (2000). *The academic achievement challenge: What really works in the classroom?* New York, NY: Guilford Publications.

Child Trends. (2015). *Data bank: Mathematics proficiency.* Bethesda, MD. Retrieved from www.childtrends.org/wp-content/uploads/2015/11/09_Mathematics_Profi ciency.pdf

Clark, K. R. (2015). The effects of the flipped model of instruction on student engagement and performance in the secondary mathematics classroom. *Journal of Educators Online, 12*(1), 91–115.

Cothran, M. (2016). Traditional vs. progressive education. Retrieved from www .memoriapress.com/articles/traditional-vs-progressive-education/

Darling-Hammond, L. (2017). *Developing and measuring higher order skills: Models for state performance assessment systems.* Washington DC: Learning Policy Institute, Council of Chief State School Officers.

Delialioğlu, Ö. (2012). Student engagement in blended learning environments with lecture-based and problem-based instructional approaches. *Journal of Educational Technology & Society, 15*(3), 310.

Desilver, D. (2017). U.S. students' academic achievement still lags that of their peers in many other countries. www.pewresearch.org/fact-tank/2017/02/15/u-s-students -internationally-math-science/

Dotterer, A. M., & Lowe, K. (2011). Classroom context, school engagement, and academic achievement in early adolescence. *Journal of Youth and Adolescence, 40*(12), 1649–1660.

Echazarra, A., Salinas, D., Méndez, I., Denis, V., & Rech, G. (2016). *How teachers teach and students learn: Successful strategies for school.* OECD Education working paper no. 130. Paris, France: Organization for Economic Co-operation and Development. Retrieved from http://www.oecd.org/officialdocuments/publicdisplaydocumentpdf /?cote=EDU/WKP(2016)4&docLanguage=En

EdSurge. (2016). *Study shows math and science scores improve for U.S. fourth and eighth grade students.* Retrieved from www.edsurge.com/news/2016–11-29-study -shows-math-and-science-scores-improve-for-u-s-forth-and-eighth-grade-students

Fennema, L., Carpenter, T., Franke, M., Levi, M., Jacobs, V., & Empson, S. (1996). A longitudinal study of learning to use children's thinking in mathematics instruction. *Journal for Research in Mathematics Education, 27*, 403–434.

Freeman, S., Eddy, S. L., McDonough, M., Smith, M. K., Okoroafor, N., Jordt, H., & Wenderoth, M. P. (2014). Active learning increases student performance in science, engineering, and mathematics. *Proceedings of the National Academy of Sciences, 111*(23), 8410–8415.

Gage, N. L. (1978). *The scientific basis of the art of teaching.* New York, NY: Teachers College Press.

Gallup (2015). *Gallup student poll—Overall report: Engage today—Ready for tomorrow.* Retrieved from www.gallupstudentpoll.com/188036/2015-gallup -student-poll-overall-report.aspx

Gningue, S. M., Peach, R., & Schroder, B. (2013). Developing effective mathematics teaching: Assessing content and pedagogical knowledge, student-centered teaching, and student engagement. *Mathematics Enthusiast, 10*(3), 621.

Gonzales, P., Guzman, J. C., Partelow, L., Pahlke, E., Jocelyn, L., Kastberg, D., & Williams, T. (2004). *Highlights from the trends in international mathematics and science study (TIMSS) 2003.* Washington, DC: National Center for Educational Statistics.

Great Schools Partnership. (2016). 21st century skills. Retrieved from http://edglossary .org/21st-century-skills/

Indiana University Newsroom. (2010). Latest HSSSE results show familiar theme: Bored, disconnected students want more from schools. Retrieved from http://news info.iu.edu/news-archive/14593.html

International Society for Technology in Education. (2017). Essential conditions: Student-centered learning. Retrieved from www.iste.org/standards/essential-con ditions/student-centered-learning

Jablon, P. (2014). *The synergy of inquiry.* Huntington Beach, CA: Shell Education.

Jenkin, M. (2013). Play in education: The role and importance of creative learning. *Guardian*, February 27. Retrieved from www.theguardian.com/teacher-network /teacher-blog/2013/feb/27/play-education-creative-learning-teachers-schools

Jimerson, S. R., Campos, E., & Greif, J. L. (2003). Toward an understanding of definitions and measures of school engagement and related terms. *California School Psychologist, 8*(1), 7–27.

Johnson, D. W., & Johnson, R. T. (2005). Training for cooperative group work. In M. A. West, D. Tjosvold, and K. G. Smith, (Eds.), *The essentials of teamworking: International perspectives* (pp. 131–147). Hoboken, NJ: John Wiley & Sons.

Kanevsky, L., & Keighley, T. (2003). To produce or not to produce? Understanding boredom and the honor in underachievement. *Roeper Review, 26*(1), 20–28.

Korb, K. A. (2011). Self-report questionnaires: Can they collect accurate information? *Journal of Educational Foundations, 1*, 5–12.

Koziol, S. M., & Burns, P. (1986). Teachers' accuracy in self-reporting about instructional practices using a focused self-report inventory. *Journal of Educational Research, 79*(4), 205–209.

Kozol, J. (2007). *Letters to a young teacher.* New York, NY: Crown.

Lamb-Sinclair, A. (2016). What if high school were more like kindergarten? *The Atlantic*. Retrieved from www.theatlantic.com/education/archive/2016/08/learning-versus-education/494660/

Larmer, J., Mergendoller, J., & Boss, S. (2015). *Setting the standard for project-based learning*. Alexandria, VA: Association for Supervision & Curriculum Development.

Lesley University NEASC Self-Study. (2015). Retrieved from www.lesley.edu/sites/default/files/2017–06/2015_Lesley_University_NEASC_Self_Study.pdf

Lieber, C. M. (2009). *Getting classroom management RIGHT: Guided discipline and personalized support in secondary schools.* Cambridge, MA: Educators for Social Responsibility.

Markham, T. (2003). *Project based learning handbook: A guide to standards-focused project based learning for middle and high school teachers*. Novato, CA: Buck Institute for Education.

Master, B., Loeb, S., & Wyckoff, J. (2017). More than content: The persistent cross-subject effects of English language arts teachers' instruction. Retrieved from *Educational Evaluation and Policy Analysis, 39*(3), 429–447. doi:0162373717691611

Mulholland, Q. (2015). The case against standardized testing. Retrieved from *Harvard Political Review*. http://harvardpolitics.com/united-states/case-standardized-testing/

Mulkey, L. M., Catsambis, S., Steelman, L. C., & Crain, R. L. (2005). The long-term effects of ability grouping in mathematics: A national investigation. *Social Psychology of Education, 8*(2), 137–177.

National Council on Teacher Quality. (2015). Policy issue: Tenure. Retrieved from https://teachertenure.procon.org/view.resource.php?resourceID = 004377

National Education Association. (2007). Jonathan Kozol: Book excerpt. *NEA Today, 26*(3), 25.

National Mathematics Advisory Panel [NMAP]. (2008). *Foundations for success: The final report of the National Mathematics Advisory Panel*. U.S. Department of Education.

Newell, R. J. (2003). *Passion for learning: How project-based learning meets the needs of 21st-century students* (vol. 3). Lanham, MD: Scarecrow Press.

Newmann, F. M. (1992). *Student engagement and achievement in American secondary schools*. New York, NY: Teachers College Press.

Nurenberg, D. (2010). A study of the effects of peaceable schools curricula on student achievement in an urban middle school. (2010). Retrieved from *Educational Studies Dissertations, 57*. http://digitalcommons.lesley.edu/education_dissertations/57

Ouyang, X. (2016). Five things to know about the PISA exam. Retrieved from *US News and World Report*. www.usnews.com/news/stem-solutions/articles/2016–11-18/5-things-to-know-about-the-pisa-exam

Owens, A., & Sunderman, G. L. (2006). School accountability under NCLB: Aid or obstacle for measuring racial equity? Retrieved from www.civilrightsproject.ucla.edu/research/k-12-education/integration-and-diversity/school-accountability-under-nclb-aid-or-obstacle-for-measuring-racial-equity/owens-school-accountability-under-nclb-2006.pdf

Panitz, T. (2000). Using cooperative learning 100% of the time in mathematics classes establishes a student-centered interactive learning environment. ERIC opinion paper # ED448063.

Partnership for 21st Century Learning. (2016). *Our vision and mission.* www.p21.org /about-us/our-mission

Prince, M. (2004). Does active learning work? A review of the research. *Journal of Engineering Education, 93*(3), 223–231.

Polly, D. (2008). Modeling the influence of calculator use and teacher effects on first grade students' mathematics achievement. *Journal of Technology in Mathematics and Science Teaching, 27*(3), 245–263.

Polly, D., McGee, J. R., Wang, C., Lambert, R. G., Pugalee, D. K., & Johnson, S. (2013). The association between teachers' beliefs, enacted practices, and student learning in mathematics. *Mathematics Educator, 22*(2), 11–30.

Polly, D., McGee, J., Wang, C., Martin, C., Lambert, R., & Pugalee, D. K. (2015). Linking professional development, teacher outcomes, and student achievement: The case of a learner-centered mathematics program for elementary school teachers. *International Journal of Educational Research, 72*(1), 26–37.

Quillen, I. (2013) Why inquiry learning is worth the trouble. Retrieved from https:// ww2.kqed.org/mindshift/2013/01/29/what-does-it-take-to-fully-embrace-inquiry -learning/

Reyes, M. R., Brackett, M. A., Rivers, S. E., White, M., & Salovey, P. (2012). Classroom emotional climate, student engagement, and academic achievement. *Journal of Educational Psychology, 104*(3), 700.

Rotherham, A. J., & Willingham, D. (2009). 21st century. *Educational Leadership, 67*(1), 16–21. Retrieved from www.ascd.org/publications/educational-leadership /sept09/vol67/num01/21st-Century-Skills@-The-Challenges-Ahead.aspx

Rowland, B. (2011). *The influence of high-stakes testing and test preparation on high school students' perspectives on education and lifelong learning.* Ann Arbor, MI: ProQuest.

Skinner, E. A., & Belmont, M. J. (1993). Motivation in the classroom: Reciprocal effects of teacher behavior and student engagement across the school year. *Journal of Educational Psychology, 85*(4), 571.

Schoenfeld, A. H. (2004). The math wars. *Educational Policy, 18*(1), 253–286.

Spring, J. (2017). *American Education.* New York, NY: Routledge.

Stephens, M., Landeros, K., Perkins, R., & Tang, J. H. (2016). *Highlights from TIMSS and TIMSS Advanced 2015: Mathematics and science achievement of US students in grades 4 and 8 and in advanced courses at the end of high school in an international context (NCES 2017–002).* Washington, DC: U.S. Department of Education, National Center for Education Statistics.

Upadyaya, K., & Salmela-Aro, K. (2013). Development of school engagement in association with academic success and well-being in varying social contexts. *European Psychologist, 18*(2), 136–147.

Valli, L., & Buese, D. (2007). The changing roles of teachers in an era of high-stakes accountability. *American Educational Research Journal, 44*(3), 519–558.

Van der Veer, R., & Valsiner, J. (1991). *Understanding Vygotsky: A quest for synthesis.* Hoboken, NJ: Blackwell Publishing.

Virtual Learning Academy. (2017). The new normal in education: How student-centered learning is transforming the classroom. Retrieved from https://vlacs.org/what-is-student-centered-learning-2/

Walker, T. (2014). NEA Survey: Nearly half of teachers consider leaving profession due to standardized testing. *NEA Today.* Retrieved from http://neatoday.org/2014/11/02/nea-survey-nearly-half-of-teachers-consider-leaving-profession-due-to-standardized-testing-2/

Wang, M. T., Brinkworth, M., & Eccles, J. (2013). Moderating effects of teacher-student relationship in adolescent trajectories of emotional and behavioral adjustment. *Developmental Psychology, 49*(4), 690.

Wang, M. T., & Holcombe, R. (2010). Adolescents' perceptions of school environment, engagement, and academic achievement in middle school. *American Educational Research Journal, 47*(3), 633–662.

Yazzie-Mintz, E., & McCormick, K. (2012). Finding the humanity in the data: Understanding, measuring, and strengthening student engagement. In J. A. Fredericks, A. L. Reschly, & S. L. Christensen (Eds.), *Handbook of research on student engagement* (pp. 743–761). New York, NY: Springer.

Yuan, K., & Le, V. N. (2012). *Estimating the percentage of students who were tested on cognitively demanding items through the state achievement tests.* Santa Monica, CA: Rand Corporation.

Chapter 7

A Phenomenological Study

Incorporating the History of Mathematics from the Perspective of Teachers

Sinem Sozen Ozdogan, Didem Akyuz, and Erdinc Cakiroglu

INTRODUCTION

Ifrah (2000) started his famous book series by providing real-life examples from his mathematics classroom in which the students were discussing popular rumors and myths related to numbers. He reflected on this lived experience in which he was unable to answer most of the students' questions, and he regretted that some of the answers that he gave them might have been wrong. The isolation of mathematics from its development process triggered him to research the phenomenon of the history of numbers.

Mathematics courses are generally not designed to incorporate the history of mathematics (Fasanelli et al., 2000; Haile, 2008); rather, teachers must enhance their instruction through the effective use of the history of mathematics in their classrooms (Siu, 2013). Thus a teacher's background knowledge is helpful for incorporating the history of mathematics into his or her lessons (Clark, 2012; Smestad, 2009). Previous studies have emphasized that although teachers tend to have a positive disposition toward incorporating the history of mathematics into their teaching, they do not prefer to use it in their classroom (Fried, 2001; Gazit, 2013; Siu, 2004).

Many studies have listed lack of time, lack of resources, and lack of teacher training as potential reasons for this reluctance (Fauvel & van Maanen, 2000; Fried, 2001; Gazit, 2013; Liu, 2003; Oprukçu-Gönülateş, 2004; Siu, 2004). Despite this, the themes gathered mostly from a literature review and survey studies with pre-service teachers or students should be evaluated and complemented with the lived experiences of in-service teachers who have worked under new policies regarding the history of mathematics. Within this perspective, there is still a need to describe the role and intentions of teachers toward incorporating the history of mathematics into an environment they experience on a daily basis.

This controversial issue is relevant to teachers in Turkey as well, and it motivated us to explore what could be the underlying reasons from the perspective of primary and middle school mathematics teachers. Although some studies have aimed to describe this issue (Fauvel & van Maanen, 2000), we are not aware of a fully phenomenological approach that aims to extract the essence of the experiences reported by primary and middle school mathematics teachers.

THE ROLE OF THE HISTORY OF MATHEMATICS IN MATHEMATICS EDUCATION

The National Council for Accreditation of Teacher Education (2003), the National Council of Teachers of Mathematics (2004), and the International Commission on Mathematical Instruction (Fauvel & van Maanen, 2000) encourage incorporating the history of mathematics into mathematics education since doing so has been found to be beneficial for both learners and teachers (Bagni, 2008; Gazit, 2013). In the literature, the main argument related to this question centers around the belief that teachers and learners can gain a deeper mathematical understanding if history is used properly in mathematics education (Bellomo & Wertheimer, 2010; Clark, 2012; Fauvel, 1991; Huntley & Flores, 2010).

The best way to present the history of mathematics in a classroom varies depending on the aim of the teaching action. Jankvist (2009) categorized these aims as history as a tool and history as a goal. History as a tool has been defined as teaching history in order to provide meaning to mathematics (Fasanelli et al., 2000). The process can be used as an affective tool, a cognitive tool, or as a tool for evolutionary arguments (Jankvist, 2009).

Using history as an affective tool pertains to the ways in which it contributes to teacher and student engagement and motivation. As a cognitive tool, history may be used to improve the teaching and learning process (Jankvist, 2009)—for example, presenting an old problem-solving strategy or providing the constitutional definition of a mathematical term may create an effective learning environment.

History as a tool for evolutionary arguments is based on the issue of recapitulation, which arises from the idea in biology that ontogeny recapitulates phylogeny (Jankvist, 2009). This helps learners understand the development of mathematical understanding that their ancestors experienced (Gulikers & Blom, 2001). The history-as-a-tool category involves incorporating the history of mathematics, which can be in the form of direct historical information from textbooks, anecdotes, visual aids, activities, role-playing, biographies, outdoor experiences, old problems, or projects (Katz, 2004; Swetz, 1994).

As for the history-as-a-goal category, if the aim is to learn the history of mathematics but not necessarily mathematics, then the history of mathematics should be considered separately from the mathematics course, which focuses on the development of mathematics itself (Jankvist, 2010). This categorization is helpful for understanding the difference between teaching history and teaching mathematics in elementary and secondary school settings.

Incorporating the history of mathematics into teaching is recommended as a classroom tool that teachers can use to support the teaching and learning of mathematics (Jankvist & Kjeldsen, 2011). Furthermore, it is argued that the history of mathematics as an educational tool should not be perceived as a separate subject; it should be infused in or should permeate the entire field of mathematics (Siu & Tzanakis, 2004).

This infusion or permeation within teaching—in other words, the incorporation of the history of mathematics—can be achieved by providing "background knowledge related to the inventor of the theory, sign, era, culture, country, in short, information that constitutes mathematics taught in the classroom to make mathematics challenging and more interesting" (Swetz, 1994, p. 2). By using historical sources, teachers and students can witness the various applications of mathematics in daily life and they can identify opportunities to investigate the mathematics of different cultures (Bagni, 2008; Grugnetti, 2000).

METHOD AND DESIGN OF THE STUDY

Like all qualitative studies, phenomenology describes human complexities; however, phenomenology can be distinguished from other approaches in terms of its profound understanding of human experiences (Padilla-Diaz, 2015). It is mainly a study of lived experience, and it seeks to directly understand the nature of experience (Moustakas, 1994). Thus human experiences are investigated using significant variables so that the essence of the experience becomes meaningful to others who encounter the same experience. *Intuition* and *intentionality* constitute the basis for a phenomenological study, as described by Moustakas, who was a follower of transcendental phenomenology.

Intuition is one way to derive knowledge about an experience free from everyday sense impressions, while intentionality refers to the conscious awareness of an experience (Moustakas, 1994). They both provide perceptions directed toward a phenomenon without judgments, biases, or everyday sense impressions. Phenomenological studies search for the meaning and essence of experience rather than try to provide an explanation of it (Moustakas, 1994; Selvi, 2008).

Transcendental phenomenology allows a researcher to develop "an objective essence through aggregating the subjective experiences of a number of individuals" (Moerer-Urdahl & Creswell, 2004, p. 23). Therefore, it is suggested that this approach can be used by studies that focus on a phenomenon in order to understand the individuals who lived or are still living the experience inherent in that phenomenon (Moerer-Urdahl & Creswell, 2004).

Selection of the Participants

In this study, the participants consisted of two primary school and two middle school mathematics teachers from four different public schools in one of the districts of Ankara, Turkey. To achieve variation in the participants, the teachers were purposefully selected based on the grade levels they taught. This contributed to variation of the data since educators teaching different grade levels present different mathematical topics and subtopics according to the national mathematics program required by the Ministry of National Education (2009).

From the 10 teachers invited to volunteer to participate in the study, four female teachers were selected. To protect their anonymity, the teachers were coded PT1, PT2, MT1, and MT2. Each teacher had at least six years of experience teaching in public schools (see Table 7.1).

Table 7.1. Profile of the Teachers in the Current Study

Name	Gender	Experience (years)	Graduation/Major	Position
PT1	Female	21	Two-year upper secondary education	Primary school teacher (third grade)
PT2	Female	16	French language teaching	Primary school teacher (fourth grade)
MT1	Female	15	Mathematics	Middle school mathematics teacher (fifth grade)
MT2	Female	6	Elementary mathematics education	Middle school mathematics teacher (sixth grade)

All of the teachers stated that they included historical information related to mathematicians in their lectures. The teachers had not attended any courses, seminars, or workshops related to the history of mathematics pre- or post-graduation. They also added that when they were students they did not have any experience of their teachers incorporating the history of mathematics in their lessons.

Field and Ethical Issues

The Middle East Technical University Ethics Committee of Graduate School of Social Sciences and the Ministry of National Education of the Turkish Republic assessed the applicability of this research. Before the study was conducted, necessary permission was obtained from these two review boards to provide that the participants were not physically or psychologically harmed during the research study (Fraenkel & Wallen, 2006). School administrators were the first to be informed of the research process.

The teachers were contacted and asked if they would like to volunteer to participate in the research study. All the participants were asked to sign a voluntary participation form. In this form, the participants were informed that they had the right to withdraw from the study at any time, and they were given complete information about the aim of the research and the data collection procedures. Before the classroom application, the participating teachers and researchers obtained verbal consent from all of the students.

Confidentiality or privacy was another concern of this research. Because each of the four participants was assigned a code based on the grade they taught, the interview protocols did not involve given names or surnames. Third-party access to the raw data was not allowed, and transcriptions from and the findings of the study were shared with each teacher for that teacher's approval.

Classroom Settings at the Macro- and Micro-level

The Higher Education Council (HEC), the institution responsible for higher education in Turkey, has stated that there should be an increase in the number of common courses and elective courses such as History of Science and Philosophy of Science in teacher education programs. The effect of this change in teacher education policy has been reflected in mathematics teacher education programs as the emergence of a course titled History of Mathematics (Higher Education Council, 2007).

The course content includes a 5,000-year history of the development of mathematics, and the course was offered at Turkish universities beginning with the 2009–2010 academic year. However, in primary teacher education programs, a History of Science course is offered instead. Similar developments have been taking place in other countries where history of mathematics courses have been integrated into teacher training programs, including those for primary school teachers (Fasanelli et al., 2000; Schubring et al., 2000).

In Turkey, the MNE incorporated the history of mathematics into primary school and middle school mathematics curricula (MNE, 2009, 2013). The results of the studies with pre-service mathematics teachers in Turkey indicated that although the teachers had a positive attitude toward this approach, they might not use it in their classrooms (Alpaslan, Işiksal, & Haser, 2014; Oprukçu-Gönülateş, 2004). The teaching behavior of the volunteers selected to participate in this present study were consistent with those reported in the literature. Additionally, through interviews and observations, it was confirmed that the teachers incorporated the history of mathematics, as they stated.

Data Collection Tools and Procedures

Each teacher was interviewed before and after the classroom observation. The topics of the week of observation were determined by the teachers based on their annual curriculum plan: conceptual understanding of fractions (third grade), calculating the perimeter of rectangular shapes (fourth grade), multiplication of four-digit numbers (fifth grade), and evaluating the similarity of triangles (sixth grade).

Prepared materials were collected from the literature related to the history of mathematics and included several types of materials such as classroom activities (see Appendix A), historical development of a topic (Appendix B), and old problems (Appendix C). These appendices provide examples of some of the materials that were prepared for the teachers. All the materials in the appendices were prepared for the teachers, not for the students. The materials given to the teachers are listed below.

PT1: The objectives of the two lesson hours focused on identifying unit fractions and exemplifying proper functions. Based on these objectives, different historical materials were prepared for PT1: direct historical information from the course books or textbooks (the meaning of fractions; broken numbers; the development process of fractions; examples from civilizations); visual aids (the Rhind Papyrus; a table of Greek numbers and a table of Egyptian numbers); old problems (a problem from papyrus); and a biography of a mathematician (Fibonacci).

PT2: The objectives of the two lesson hours focused on measuring the circumferences of geometric shapes and identifying the relationship between the circumference and the side length of squares and rectangles. Based on these objectives, different historical materials were provided for PT2: direct historical information from course books or textbooks (the meaning of geometry; the development of measurement and ancient Egypt); visual aids (an example of papyrus; a video about the River Nile); old problems (a problem from papyrus); and the biography of a mathematician (the contributions of Atatürk to geometry).

MT1: The objectives of the two lesson hours focused on doing multiplications that resulted in, at most, seven-digit products. Based on these objectives, different historical materials were provided for MT1: direct historical information from course books or textbooks (the development of multiplication and division; where did the four different multiplication symbols come from?); visual aids (an example of old mathematics books); activities (the Italian multiplication method, the Gelosia method, and Rabdologia); and the biography of a mathematician (John Napier).

MT2: The objectives of the two lesson hours focused on guessing and measuring the circumferences of geometric shapes. Based on these objectives, different historical materials were provided for MT2: direct historical information from course books or textbooks (the meaning of geometry; the development process of standardized measurement units; ancient Egypt); visual aids (an example of papyrus); anecdotes (King Henry I); old problems (a problem from China); and biographies of mathematicians (the contributions of Atatürk to geometry; Euclid).

Interviews

Each participant attended three face-to-face and semi-structured interviews during the study. All the interviews were conducted on different days. The first interview included two parts. The first part consisted of general questions related to information about the teacher's personal background and personal experiences of teaching mathematics. The second part consisted of questions about the teacher's past experiences related to the history of mathematics. Therefore, the first interview served as a warm-up to acknowledge the teacher as a participant and her experience of teaching mathematics and the history of mathematics.

As it was important to ensure that the teachers were conscientious in incorporating the history of mathematics (Moustakas, 1994), they were provided with historical materials that included direct historical information from books, anecdotes, visual aids, activities, biographies, and old mathematical problems. One of the researchers prepared these materials for each teacher separately in the form of various materials.

The teachers were also encouraged to incorporate any historical information relevant to the topic from reliable sources that they found. They were also informed that they could use the historical material partially or in its entirety as long as they needed it for the sake of teaching and learning so that the history of mathematics permeated the teaching process when it was needed as a tool for teaching and learning.

In the second interview, the participants reflected on these historical materials after studying and incorporating them in their lesson plans. In the last interview, the participants described their experience after presenting two lesson hours of classroom experience in detail. The first question focused on how the teachers described their experience of incorporating the history of mathematics into teaching. The other questions evolved throughout the third interview to investigate emerging phenomena—for example, "Why did you prefer using specific historical material, for example, direct historical knowledge?" or "Can you give an example from the classroom experience?"

Observation

Each teacher was observed for two lesson hours to better understand the descriptions of the classroom experiences they provided in their interviews. As a nonparticipant observer, the researcher did not interfere with the flow of the instruction. The teachers incorporated at least one of the historical materials into their teaching partially or in its entirety.

The primary school teachers offered to divide the two-hour classroom observation into two parts due to their usual schedule. They avoided teaching two successive lesson hours of mathematics because their young students had a limited ability to concentrate. The classroom practices of the teachers were videotaped while the researcher took notes. Transcriptions of the observations were used to support and enrich the data collected from the interviews to ensure the credibility of the inferences that the researcher made (Moerer-Urdahl & Creswell, 2004).

Analysis

Data analysis is considered a difficult part of transcendental phenomenology (Moerer-Urdahl, 2004). In this study, the Stevick-Collaizzi-Keen model modified by Moustakas (1994) was used as a guide to analyze the data. *Epoché*, or bracketing, in a phenomenological study allows the researcher to be aware of his or her personal assumptions about and biases toward the phenomenon in order not to affect the subjectivity of the participants (Moustakas, 1994). Thus the researcher who conducted the interviews bracketed her assumptions about and biases toward incorporating the history of mathematics into the teaching approach.

From the beginning of the study, the researcher was conscious of her tendency and dedicated herself to bracketing her personal biases or prejudgments by avoiding explanations and using three interviews to make the participants feel comfortable about sharing their experiences based on their own intentionality and intuition. All the data gathered from the interviews and observations were transcribed.

The transcriptions were checked for accuracy by re-reading them until every part was clearly understood. The data collected from the interviews were reported thoroughly to explain the incorporation of the history of mathematics into mathematics teaching. Significant statements, or horizons, were selected from the transcriptions by setting aside irrelevant, repetitive, or overlapping segments. This is known as *phenomenological reduction* (Moustakas, 1994).

Once recorded, these statements constituted a description of the experience and are referred to as *textural descriptions*. At this point, *imaginative variation* enables the researcher to grasp the structural descriptions of the experience (Moustakas, 1994). The term *textural description* refers to what is being described. The term *structural description* refers to how something is expressed by the participants (Padilla-Diaz, 2015).

Table 7.2 shows the synthesis of the textural and structural descriptions that give the essence of the experience in a particular time and place (Moerer-Urdahl & Creswell, 2004; Moustakas, 1994) as provided in the findings section of the study.

Table 7.2. Examples of the Composite Textural and Structural Descriptions

Themes	Textural Descriptions	Structural Descriptions
Professional Development	"It [the approach] is good for me as well. I learned something about the history of mathematics. It also made me think that it made contributions to my knowledge."	Developing content knowledge
	"Why would we feel confined to use the knowledge that we already knew? There is no end for mathematics and for other disciplines as well."	Developing content knowledge and gaining insight about mathematics
	"Mathematics is a product of human efforts."	Gaining insight about mathematics
	"Mathematics is not something that emerged on its own."	Gaining insight about mathematics
	"I realized that there existed great effort and surmised that it would be in the future very beneficial for us or for the next generations."	Gaining insight about mathematics

Findings

The findings of the study demonstrated that the teachers described and perceived using history using eight core themes: (1) professional development, (2) attractiveness, (3) an environment for speaking up, (4) the need for experience, (5) adaptability of historical materials, (6) problem solving versus incorporating the history of mathematics, (7) lack of knowledge and resources, and (8) time constraints.

Professional Improvement

The teachers mentioned some benefits of teaching the history of mathematics for themselves. The participants evaluated these benefits considering their knowledge related to mathematical content. They elaborated this issue and added that the history of mathematics helped them gain insight into mathematics. The teachers said that while they were studying the historical materials, the process provided general knowledge about mathematics that they considered knowledge a teacher or student should have. Additionally, MT1 regarded it as the "celebrity news of mathematics" since the history of mathematics included the life of mathematicians and anecdotes. PT2 and MT1 thought that the historical information contributed to their mathematical knowledge because it helped them answer the "Why?" questions of students. All of the teachers agreed that while studying the history of mathematics, they pushed themselves to think about the existence of mathematics. PT1 said,

> I realized that there existed great effort [in mathematics as a collective activity] and surmised that it would be very beneficial for us in the future or for the next generations. I gained insight that next generations would benefit from us, add something to it [mathematics].

The teachers concluded that "now" was not an end for mathematics because mathematics continues to develop. They deduced that methods, concepts, formulas, and definitions may change as long as people study mathematics, just like anything in life. MT1 gave an example: "You use the best car model but . . . three years later your car will be out of fashion. However, it is in our power to change it."

Attractiveness

Attractiveness was described by the teachers as an aspect of incorporating the history of mathematics into teaching, and there were times in the classroom when both the students and the teacher enjoyed themselves while discussing historical materials. The teachers further explained that each mathematics lesson might not be equally attractive to both students and teacher. They attributed this to three reasons: uninterested students, uninteresting subjects, or uninteresting methods of teaching.

Maintaining the students' attention was regarded as an important issue for the teachers. For this reason, the teachers frequently emphasized it during their interviews. According to them, the indifference of the students resulted from the "information age." They explained that it was easy for students to acquire information willingly or unwillingly. The teachers added that students became uninterested in topics because students thought that they had already heard or knew the mathematical knowledge presented in the classroom. The teachers also agreed that being presented with similar topics year after year, albeit at different levels of difficulty, would be boring for the students.

At that point, the teachers complained about not being able to maintain enough student attention to help them focus on the lesson. Even so, the teachers viewed incorporating the history of mathematics into teaching as interesting and believed that it could improve their ability to keep the students' attention. The teachers also thought that their students were interested in the historical materials:

> When talking with my students, I would understand from their facial expressions if they were bored. However, this lesson was never found to be boring. I even wanted to solve the [historical] problem myself at first. It [the problem] drew my attention as well. Solving those questions immediately, I found myself saying, "Let's solve it. What might be the answer?"

Moreover, PT1 said that "although the children are quite young, talking about papyrus or the Egyptians got them interested in the topic." Similar examples relevant to the attractiveness of the historical materials were given by the other teachers. In addition, enjoyment was another factor that made the history of mathematics more attractive. The teachers agreed that they enjoyed teaching the history as much as students enjoyed learning it. MT1 commented, "I don't think that they will use it in daily life, however, the method [the Gelosia method for multiplication] was enjoyable."

An Environment for Speaking Up

The teachers pointed out that the history of mathematics incorporated lesson hours within the mathematics lesson that allowed students to "speak up." Within this framework, the "speaking-up" environment was evaluated by the teachers under two phases: how students discussed their ideas about mathematics and how actively they participated in the class. First, the teachers commented that "unexpected" students found the chance to express themselves in the discussion environment. MT2 expressed her experience this way:

> In the lesson, students participated as usual. However, it [the lesson] was different than usual since different students participated. I mean, a discussion environment does not happen all the time; it was the first time that a further discussion environment was created.

The teachers stated that this discussion environment was triggered by the challenging elements provided by the history of mathematics. According to the teachers, the students were thinking deeply about mathematics. They questioned what they were learning and this led to challenging discussions between students and the teacher.

The teachers emphasized the importance of questioning mathematics using examples from the classroom. They said that a challenging discussion environment was a way of questioning mathematics. It was also emphasized that questioning was necessary for the classroom environment. MT1 added that the teacher's role is to make students question mathematics:

> Students may not question, it was again up to me to provide them with a stimulation for questioning, but first I need to wonder so that I can make my students wonder. [If I don't] it is my incompetency then."

As for the second phase, the teachers commented that the students actively participated in the classroom activities during these challenging discussions. It was the teachers' complaint that direct mathematical instruction did not provide an environment in which students wanted to participate. The teachers said that this new environment motivated the students to speak up. The teachers explained that when they incorporated the history of mathematics in their lesson, the participation in the classroom environment was high and even surpassed their expectations. Moreover, as MT2 explained,

> normally, if 7 to 8 students participate in solving the problems because the problem is hard [for the other students], of course the number of raised hands to open-ended questions will be greater. Therefore, we enabled more students to participate.

The active participation of the students, a challenging discussion environment, and questioning constituted a speaking-up environment that the teachers found to be motivating. The teachers indicated that students expressed their ideas freely in a speaking-up environment and were motivated to participate actively in the learning process.

The Need for Experience

The teachers believed that incorporating the history of mathematics in their lessons was an approach that improved with experience. They described themselves as nervous and unsuccessful during the first hour of observation. During the second hour, when they were more experienced, they felt that the students were also more interested. The teachers discussed the idea that students and teachers needed practice to "adapt" to incorporating the history of mathematics and to perform better. PT2 commented,

> In the first lesson I was more stressed "What can I do? How can I do it?" but in the second lesson I was much more flexible.

Three of the teachers stated that they incorporated the history of mathematics more than once after the classroom observation ended to test their belief that "the more practice there is, the greater the success" before classroom observation with other classes.

MT1 and MT2 found a chance to incorporate the same materials into other classrooms while PT1 shared the experience of using the same materials in her third-grade classroom for the second time and getting different and interesting answers to the same questions from her students. Since the teachers were satisfied with the results, they concluded that they would continue to incorporate the historical material given.

> We didn't use this method within the lesson but I am sure that I will again talk about it [the material provided for the study] in order to draw students' attention. I am sure that I will discuss it.
>
> It is a method [the Gelosia method] that I will add to my next fifth-grade program. Although it will not cover two hours, in one lesson hour I can help the students practice.
>
> As long as we discuss these topics [history of mathematics materials about fractions], maybe as long as we incorporate the history of mathematics not only with fractions but with every topic, it will be more beneficial, I think, and I will want to use it in my lessons.

The teachers said that they would continue to incorporate the historical materials given. One of the teachers even added that she would incorporate it in every mathematical topic. The teachers were in agreement that using the given historical materials during lessons was beneficial and they would be willing to incorporate them in other lessons.

The Adaptability of Historical Materials

The teachers said that they had some concerns about how to present the historical material in the classroom. Their concern was in regard to the adaptability of the historical materials to the objectives (learning outcomes) of the mathematical topic and each particular age group. The teachers were asked whether incorporating the history of mathematics left them behind in terms of their objectives, and they said this depended on the objectives they were teaching.

For this study, the materials were found to be appropriate to the topic they were teaching, but they agreed that some of the objectives they were teaching might make incorporating the history of mathematics more difficult. These objectives are seen as important for classroom practices. According to the teachers, it was not possible for all the students in the classroom to reach a learning objective at the same time. Therefore, they said, falling behind in their program was not the main concern but rather maintaining consistency between the objective and the historical material.

Another concern that the teachers frequently mentioned was the appropriateness of the historical materials for each student age group. They believed that the age of the students would be a condition for students to understand historical information. The teachers described this concern in terms of two aspects. One of them was the "prior knowledge of the students." The teachers believed it was important to decide which historical materials were going to be presented in the classroom. The prior knowledge of the students was relevant to their knowledge of other disciplines and their current mathematical knowledge. Two examples given by the teachers were the following:

> Students do not know Archimedes from science lessons, I suppose, therefore I cannot mention Archimedes in the lesson. (MT2)

> I consider the age group. For example, the type of multiplication symbol to use. We will talk about them when the time comes. I do not want to tire them. We just simply use x now. Therefore, I do not mention the other symbols. (MT1)

The teachers used the word "eliminate" when deciding on appropriate historical materials according to the age group and the objectives of the topic. According to them, they eliminated inappropriate materials by considering the age group and the objectives of the topic.

After elimination, materials that were considered appropriate for the classroom were selected according to their different properties. The teachers insisted on incorporating "simple" information. They described simplicity using terms such as *not complex*, *not comprehensive*, *not very detailed*, *concrete*, *shallow*, and *not boring*.

Furthermore, the teachers selected "remarkable" information from the historical materials, information that did not require rote learning or memorization. For them, memorizable information consisted of different civilizations, the meaning of mathematical terms, different methods, or answers to the questions Who did it? What did they do? When did they do it? PT2 also added that there should not be all the biographical information of the mathematicians. There should be less information, and the remaining information should be important and relevant. They confirmed the "remarkableness" of the information by asking students questions about what they had been taught in the previous lesson hour.

Generally, each teacher found it possible to incorporate the history of mathematics in their teaching by first considering the age group, their own goals, and their objectives.

Problem Solving vs. Incorporating the History of Mathematics

The math teachers said that they always tried to allocate time for problem solving in each mathematics lesson. The teachers defined problem solving as a tool for understanding mathematics. MT1 stated, "I think students understand better the topic when we solve problems." The teachers noted that in their lesson plans there were problems that "absolutely" had to be solved—for example, one teacher said, "After discussing a subject, I know we need to solve three problems related to the subject, and that's how I prepare my plan."

The teachers were very uncertain whether there was a benefit in dedicating some of their allocated problem-solving time to the history of mathematics. This led them to a dilemma: problem solving or the history of mathematics. The importance of problem solving was described in different ways: their students might demand to solve problems, for example, because otherwise they would have an insufficient command of the content. Moreover, the teachers believed that "the more problems that are solved, the more successful the lesson becomes." One teacher described her concern this way:

> For me, on the one hand, I was constantly worried that I needed to teach the subject and after that solve the problems as well. I cannot help it but I tend to solve problems because if I do not, I feel like the students will not be able to solve problems in the examination. On the other hand, the reference to history—this aspect is what I enjoyed. I am just saying, does achieving the objective mean solving problems? No. However, if we do not, we feel we have not done as much as we can.

One of the teachers also said that problem solving was not a skill that everyone had but without problem solving, a student could not improve him- or herself. According to the teachers, historical information and problem solving were not interchangeable in terms of time and skills. They all said that solving fewer problems made them worry, and this situation could not be "repaired" by teaching the history of mathematics.

Lack of Knowledge and Resources

When the teachers were given historical materials, they said their first thought was "I need to research this." Their explanation was that they needed to question the accuracy of what was written in the historical materials, they felt inadequate to do this, and they wondered, "Where can I get information I need?" The second question of the teachers was "How do I decide which historical information to use and where to incorporate it?" PT1 said,

> Indeed, I knew I wanted more information. I searched on the Internet but I could not find much more information.

This sense of a lack of resources was felt by each teacher. They felt uncomfortable "not having a grasp of the subject." Two of the teachers (PT2 and MT2) said that they had to know their students' many conceptions and misconceptions so that they could help their students question what they knew.

Therefore, the teachers said that because they were new to the approach they could not "go any further than the historical materials." As a result, the historical materials did not satisfy them; on the contrary, their feeling of insufficiency in historical knowledge increased:

> Perhaps, if I had a grasp of the [historical] subjects, it would be more beneficial. I cannot say it was useless. It was beneficial but it was not a 100% efficient lesson. If I had a grasp of the subject, maybe, time management would be much better. The hints I gave for the questions could be much better, maybe.

This lack of knowledge seemed to make them feel insufficient in terms of incorporating the historical materials. MT1 even associated her reason for not incorporating such materials before with her lack of knowledge. She added that her lack of knowledge originated from both her secondary and undergraduate education. Indeed, none of the teachers had been educated with a historical approach in mathematics. Therefore, they did not intuitively grasp the history of mathematics approach. MT1 expressed her feelings this way:

> If I knew the studies related to history of mathematics, when appropriate I would talk to students about those studies . . . Even if I did not show any method, just giving a piece of information would be great, I think.

MT1 added that if she'd known the historical information of a topic before, she would have shared it with the students. The teachers experienced this feeling related to the lack of historical knowledge throughout the study, and they confessed that the feeling continued to exist after the study ended. PT2 and MT2 asked the researcher for extra materials for further classroom experiences such as books or websites about the history of mathematics.

MT2 gave an example of biographical information of mathematicians written in the teachers' handbook provided by the Turkish government. She stated that the content was not "interesting to read"—for example, raw data such as the mathematician's date of birth and death. All the teachers agreed that visual materials should be available for the sake of mathematics education such as "presentations" or "cartoons" related to the history of mathematics. All in all, they needed to be trained in incorporating the history of mathematics in their lesson plans.

Time Constraints

Time constraints were mentioned as a common problem for the teachers. Four lesson-hours per week are allocated for mathematics education at both the primary and middle grade level. All of the teachers did not find this time sufficient to complete the annual mathematics program. They added that they could not even do the activities or complete the exercises in the students' textbook. Therefore the teachers always felt constrained by time. PT1 said,

> Four hours per week is very limited. For this reason, I do not think we teach mathematics adequately. It is not productive due to the lack of time.

Due to these time constraints, teachers approach novelties within teaching cautiously, and this is what initially happened when they were asked to incorporate the history of mathematics in their lessons. PT2 commented,

> I was opposed to the history of mathematics before, but if we manage the time well, we can incorporate it within the lessons, I think.

Time management was important to them in handling historical materials, solving problems, and discussing each subject. The teachers emphasized that when they had experience in the approach, they began to manage their time better.

DISCUSSION AND IMPLICATIONS

Eight themes of the teachers' experiences emerged from the data (see Figure 7.1). The teachers described incorporating the history of mathematics in their teaching as a means of professional improvement; they also mentioned the attractiveness of the historical materials for creating an environment in which their students felt free to speak up. The recent literature has shown the effect of incorporating the history of mathematics on teachers' content knowledge and pedagogical content knowledge (Clark, 2012; Huntley & Flores, 2010; Mosvold, Jacobsen, & Jankvist, 2014).

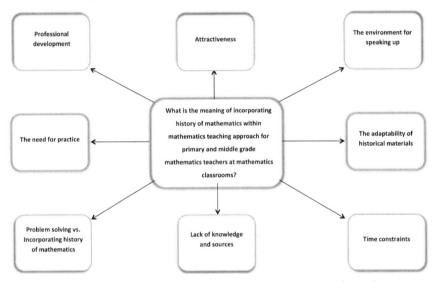

Figure 7.1. Themes revealed by the teachers' lived experiences. *Authors*: Sinem Sozen Ozdogan, Didem Akyuz, and Erdinc Cakiroglu

In this study, the teachers regarded historical information as contributing to the general mathematical knowledge that a student or a teacher should know, and they found it beneficial in answering the students' "Why?" questions. NCTM (2004) noted that chronological questions can be best explained by the history of mathematics. In addition, Mosvold et al. (2014) provided examples to show how mathematical knowledge for teaching can be developed by using the history of mathematics. However, the participants of this study did not address this issue in their experience. This theme suggests the need to design in-service teacher preparation programs both to keep up with curriculum changes and to integrate the practical experiences of each teacher using a historical perspective (Clark, 2012),

The teachers in this study asserted that most of the students actively participated when the teachers incorporated the history of mathematics in their lessons. This finding is similar to results reported by Gulikers and Blom (2001). During the history-of-mathematics discussions, different students had an opportunity to express themselves with open-ended questions. Thus the attractiveness of the material and the creation of an environment in which students were able to "speak up" allowed these students to actively participate in the classroom discussions and activities.

However, the teachers had three constraints that worried them: lack of knowledge and resources; time limitations; and problem solving versus incorporating the history of mathematics. Time constraints have been discussed in the literature (Fried, 2001; Siu, 2004; Panasuk & Bolinger-Horton, 2012; Tzanakis & Arcavi, 2000). The participants in this study believed that lack of time always creates a problem for incorporating something new in the mathematics curriculum, which leads to problems in time-management.

As a solution, they emphasized the need for greater experience in the incorporation of historical materials. The more they could practice using the material, the better they became. Lack of knowledge or resources could be understood as the major problem associated with this experience, as Fauvel (1991) mentioned. Although technological advances have become an indispensable part of education, this problem still continues 20 years later (Siu, 2013). When time constraints are viewed through the lens of professional improvement, teachers do not have the background knowledge they need to efficiently incorporate the history of mathematics in their lessons (Alpaslan, et al., 2014; Haile, 2008; Panasuk & Bolinger-Horton, 2012; Siu, 2004).

For example, to find new resources the teachers searched the Internet first and then they searched their mathematics textbooks; however, they were unable to satisfy their own learning needs at first. From a broader perspective, this could be a problem in countries whose official language is not English (the teachers' mother tongue was Turkish and they did not have a second language). This issue could be further investigated in future studies. Although the teachers had doubts, they seemed determined to continue to actively incorporate the history of mathematics by trying to identify more sources, requesting the materials provided to the other teachers and offering their materials to pre-service teachers.

Another constraint that worried the teachers was time allocation in regard to solving problems versus incorporating an understanding of mathematics history into the lesson. Although some of the teachers provided historical problems in the classroom, they still regarded this as a concern. For many teachers, historical problems might not have the same meaning as the mathematical problems they routinely solve in their classrooms. A similar concern was reported by Lim and Chapman (2015). In that study, the students asked to spend more time on exam-related materials and tutorial questions.

Cultural issues such as national examinations or grade-point average may affect the demands placed on teachers and learners, thus impacting the flow of a course. As Meavilla and Flores (2007) noted, a course design including old mathematical problems from different civilizations might help teachers see how to integrate problems and mathematical history.

Finally, the teachers of the study proposed solutions for improving their classroom practices of incorporating the history of mathematics by focusing on the adaptability of the historical materials and satisfying the need for their own experience in the material. They viewed lack of knowledge and resources as constraints of their professional improvement. If the teachers did not encounter any problems with the resources, the adaptability of the historical materials would be a concern, not a constraint. Schubring et al. (2000) discussed the concerns of pre-service and in-service teachers, and one of them was the need for more-suitable teaching materials.

In this study, the teachers adapted the materials given to them based on their own objectives and their knowledge of the students' prior knowledge of mathematics and other disciplines. Visual and hands-on materials may be prepared for teachers, as was done for the participants in this study. While institutions and organizations are in favor of incorporating the history of mathematics into the curricula (Fasanelli et al., 2000; Schubring et al., 2000), teachers should be encouraged to achieve that goal.

The need for experience in the material recommends incorporating the history of mathematics into training programs for in-service and pre-service teachers. The methodological approach used in this study might be beneficial for developing the intentionality of teachers with an appropriate adaptation to teacher education programs.

CONCLUSION

This study may provide clues about the possible common and distinct ideas underlying themes related to teachers' perceptions of using the history of mathematics while teaching mathematics. The main contribution of the study is that it gathered themes perceived by the teachers as being related to the essence of their experience of incorporating the history of mathematics into their lessons so that the experience would not be thought of as something separate and apart from their teaching; rather, it was seen as being a meaningful part of it.

This study's results strengthened findings reported in previous studies, and it elaborated upon and emphasized some of the concerns that still need to be considered in both pre-service and in-service teacher training programs. There was no specific difference between the primary school teachers' experiences and the middle school mathematics teachers' experiences. Any noted differences resulted from their distinct backgrounds.

Although studies related to primary school teachers' experience of incorporating the history of mathematics into their lessons are rather limited, it was enlightening to find common themes based on the perceptions of primary school teachers and middle school mathematics teachers. This data should facilitate our ability to take action to enhance the education of primary school teachers as well.

APPENDICES

Appendix A

Multiplication with Gelosia Method

Lesson: Mathematics
Grade Level: 5th grade (10–12 year-old)
Learning Area: Numbers
Topic: Multiplication of numbers with three or more digits

Activity Time

There have been several multiplication methods throughout its history. Gelosia method is one of the old multiplication algorithms popular among nations in medieval times. Let's understand the method and discuss its opportunities or difficulties.

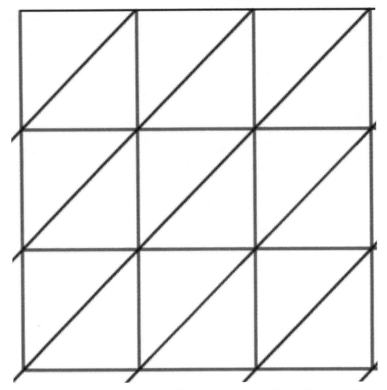

Figure 7.2. Represents multiplication of two three-digit numbers.
Authors: Sinem Sozen Ozdogan, Didem Akyuz, and Erdinc Cakiroglu

The figure above consists of three rows and three columns which represent multiplication of two three-digit numbers. There are squares showing the intersection of rows and columns; and they are divided into two diagonally. Why?

Let's try to do a first example; 372 × 431 = ?

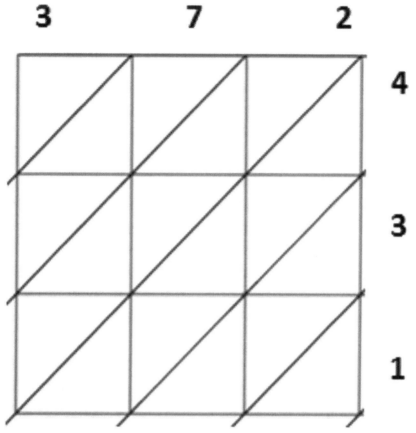

Figure 7.3. Each cell represents the tens digit. *Authors*: Sinem Sozen Ozdogan, Didem Akyuz, and Erdinc Cakiroglu

Upper piece in each cell represents the tens digit and the lower piece is the ones digit.

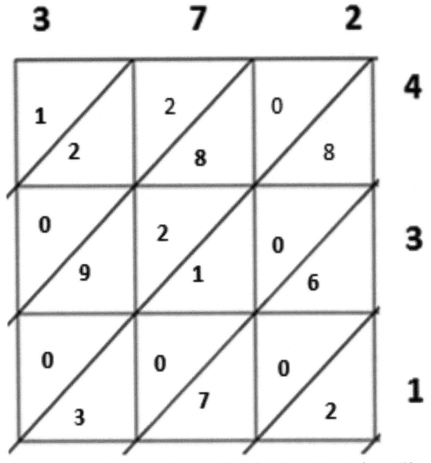

Figure 7.4. Each cell represents the ones digit. *Authors*: Sinem Sozen Ozdogan, Didem Akyuz, and Erdinc Cakiroglu

After we write the multiplication of each row and column in the cells, we calculate the total number diagonally as shown below:

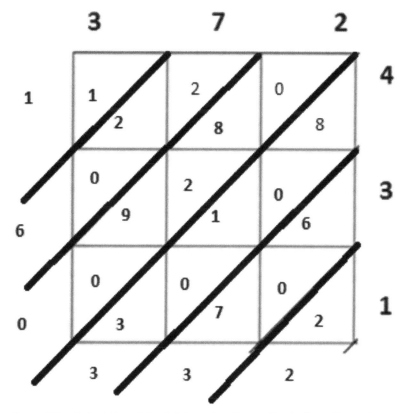

Figure 7.5. Calculation of the total number diagonally. *Authors:* Sinem Sozen Ozdogan, Didem Akyuz, and Erdinc Cakiroglu

The result is 160,332, reading from up to down, and from left to right.

Further Questions

1. What about multiplication of four-digit numbers? Does this algorithm work?
2. What about multiplication of two-digit numbers with four-digit numbers such as 24 × 5,698? Does this method still work?
3. What are the difficulties or advantages of this method?

APPENDIX B

A PERSPECTIVE TO THE HISTORICAL DEVELOPMENT OF FRACTIONS[1]

Lesson: Mathematics
Grade Level: 3rd grade (9–11 year-old)
Learning Area: Numbers
Topic: Conceptual understanding of the fractions

While fractions occur in the oldest mathematical records which have been found, the ancients nevertheless attained little proficiency in them. Evidently this subject was one of great difficulty. Simultaneous changes in both numerator and denominator were usually avoided. Fractions are found among the Babylonians. Not only had they sexagesimal divisions of weights and measures, but also sexagesimal fractions. These fractions had a constant denominator (60), and were indicated by writing the numerator a little to the right of the ordinary position for a word or number, the denominator being understood. We shall see that the Romans likewise usually kept the denominators. Ahmes confines himself to fractions of a special class, namely *unit-fractions,* having unity for their numerators. A fraction was designated by writing the denominator and then placing over it either a dot or a symbol, called *ro.*

Figure 7.6. Explaining the origin of a fraction. *Authors*: Sinem Sozen Ozdogan, Didem Akyuz, and Erdinc Cakiroglu

The big part in the picture above is called ro. Fractional values which could not be expressed by any one unit-fraction were represented by the sum of the two or more of them. Thus he wrote $\frac{1}{6}$ $\frac{1}{18}$ in place of $\frac{2}{9}$.

APPENDIX C

OLD PROBLEM FROM CHINA[2]

Lesson: Mathematics
Grade Level: 6th grade (11–13 year-old)
Learning Area: Geometry
Topic: Similarity

This problem was found in the Chinese documents from 100 B.C, nine chapters *pn* the mathematical art. In this document, there are 246 problems and their solutions. The problems all pertained to the bureaucratic needs of the Chinese empire and concern such topics as the distribution of grain, the collection of taxes, and the construction of dikes and fortifications. The problems are divided into nine chapters, with each chapter focusing on a specific application. One of the problems:

A square-walled city of unknown dimensions has four Gates, one at the center of each side. A tree stands 20 *pu* from the North gate. One must walk 14 *pu* southward from the South gate and then turn West and walk 1,775 *pu* before he can see the tree. What are the dimensions of the city?

The answer: Each side is 250 pu long.

NOTES

1. Cajori, F. (1896). *A history of elementary mathematics, with hints on methods of teaching.* London: The Macmillion Company.
2. Swetz, F. J. (1994). *Learning activities from the history of mathematics.* Portland, ME: Walch Publishing.

REFERENCES

Alpaslan, M., Işiksal, M., & Haser, Ç. (2014). Pre-service mathematics teachers' knowledge of history of mathematics and their attitudes and beliefs towards using history of mathematics in mathematics education. *Science & Education, 23*(1), 159–183. doi:10.1007/s11191-013-9650-1

Bagni, G. T. (2008). A theorem and its different proofs: History, mathematics education, and the semiotic-cultural perspective. *Canadian Journal of Science, Mathematics, and Technology, 8*(3), 217–232.

Bellomo, C., & Wertheimer, C. (2010). A discussion and experiment on incorporating history into the mathematics classroom. *Journal of College Teaching & Learning, 7*(4), 19–24.

Clark, K. M. (2012). History of mathematics: Illuminating understanding of school mathematics concepts for prospective mathematics teachers. *Educational Studies in Mathematics, 81*(1), 67–84. doi:10.1007/s10649-011-9361-y

Fasanelli, F., Arcavi, A., Bekken, O., Carvalho e Silva, J., Daniel, C., Furinghetti, F., Grugnetti, G., Hodgson, B., Jones, L., Kahane, J.-P., Kronfellner, M., Lakoma, E., van Maanen, J., Michel-Pajus, A., Millman, R., Nagaoka, R., Niss, M., Pitombeira de Carvalho, J., Silva da Silva, C. M., Smid, H. J., Thomaidis J., Tzanakis, C., Visokolskis, S., & Zhang, Dian Zhou, C. (2000). The political context. In J. Fauvel & J. van Maanen (Eds.), *History in mathematics education: The ICMI study* (pp. 1–38). Dordrecht, The Netherlands: Kluwer Academic.

Fauvel, J. (1991). Using history in mathematics education. *For the Learning of Mathematics, 11*(2), 3–6.

Fauvel, J., & van Maanen, J. (Eds.). (2000). *History in mathematics education: The ICMI study*. Dordrecht, The Netherlands: Kluwer Academic.

Fraenkel, J. R., & Wallen, N. E. (2006). *How to design and evaluate research in education*. New York, NY: McGraw-Hill.

Fried, M. N. (2001). Can mathematics education and history of mathematics coexist? *Science and Education, 10*(4), 391–408.

Gazit, A. (2013). What do mathematics teachers and teacher trainees know about the history of mathematics? *International Journal of Mathematical Education in Science and Technology, 44*(5), 501–512. doi:10.1080/0020739X.2012.742.151

Grugnetti, L. (2000). The history of mathematics and its influence on pedagogical problems. In V. J. Katz (Ed.), *Using history to teach mathematics: An international perspective* (pp. 29–35). Washington, DC: Mathematical Association of America.

Gulikers, I., & Blom, K. (2001). "A historical angle": A survey of recent literature on the use and value of history in geometrical education. *Educational Studies in Mathematics, 47*, 223–258.

Haile, T. K. (2008). *A study of the use of history in middle school mathematics: The case of connected mathematics curriculum* (Unpublished doctoral dissertation). University of Texas, Austin, Texas.

Higher Education Council [HEC]. (2007). *Öğretmen yetiştirme ve eğitim fakülteleri: 1982–2007* [Teacher training and faculties of education: 1982–2007]. Ankara, Turkey: Meteksan.

Huntley, M. A., & Flores, A. (2010). A history of mathematics to develop prospective secondary mathematics teachers' knowledge for teaching. *PRIMUS: Problems, Resources, and Issues in Mathematics Undergraduate Studies, 20*(7), 603–616.

Ifrah, G. (2000). *The universal history of numbers: From prehistory to the invention of the computer.* New York, NY: John Wiley & Sons.

Jankvist, U. T. (2009). A categorization of the "whys" and "hows" of using history in mathematics education. *Educational Studies in Mathematics, 71*, 235–261.

Jankvist, U. T. (2010). An empirical study of using history as a "goal." *Educational Studies in Mathematics, 74*, 53–74. doi:10.1007/s10649-009-9227-8

Jankvist, U., & Kjeldsen, T. (2011). New avenues for history in mathematics education: Mathematical competencies and anchoring. *Science & Education, 20*(9), 831–862. doi:10.1007/s11191-010-9315-2

Katz, V. J. (2004). *The history of mathematics: Brief edition.* Irving, TX: Pearson Education.

Lim, S. Y., & Chapman, E. (2015). Effects of using history as a tool to teach mathematics on students' attitudes, anxiety, motivation and achievement in grade 11 classrooms. *Educational Studies in Mathematics, 2*, 189. doi:10.1007/s10649-015-9620-4

Liu, P-H. (2003). Do teachers need to incorporate the history of mathematics in their teaching? *Mathematics Teacher, 96*(6), 416–421.

Meavilla, V., & Flores, A. (2007). History of mathematics and problem solving: A teaching suggestion. *International Journal of Mathematical Education in Science and Technology, 38*(2), 253–259.

Ministry of National Education [MNE]. (2009). *İlköğretim matematik dersi 1–5. sınıflar öğretim programı* [Primary mathematics program for grades 1–5]. Ankara: Milli Eğitim Basımevi.

Ministry of National Education [MNE]. (2013). *Ortaokul matematik dersi (5, 6, 7 ve 8. sınıflar) öğretim programı* [Mathematics curricula program for grades 5, 6, 7, and 8]. Retrieved from http://ttkb.meb.gov.tr/www/guncellenen-ogretim-programlari-ve-kurul-kararlari/icerik/150

Moerer-Urdahl, T., & Creswell, J. (2004). Using transcendental phenomenology to explore the "ripple effect" in a leadership mentoring program. *International Journal of Qualitative Methods, 3*(2), 1–28. Retrieved from www.ualberta.ca/~iiqm/backissues/3_2/pdf/moerercreswell.pdf

Mosvold, R., Jakobsen, A., & Jankvist, U. (2014). How mathematical knowledge for teaching may profit from the study of history of mathematics. *Science & Education, 23*(1), 47–60. doi:10.1007/s11191-013-9612-7

Moustakas, C. (1994). *Phenomenological research methods.* Thousand Oaks, CA: Sage Publications.

National Council for Accreditation of Teacher Education [NCATE]. (2003). NCATE/NCTM program standards (2003): Programs for initial preparation of mathematics teachers. Retrieved from www.ncate.org/LinkClick.aspx?fileticket=ePLYvZRCuLg%3D&tabid=676

National Council of Teachers of Mathematics [NCTM]. (2004). *Historical topics for the mathematics classroom* (2nd ed.). Reston, VA: Author.

Oprukçu-Gönülateş, F. (2004). *Prospective teachers' views on the integration of history of mathematics in mathematics courses.* (Unpublished master's thesis). Boğaziçi University, İstanbul, Turkey.

Padilla-Diaz, M. (2015). Phenomenology in educational qualitative research: Philosophy as science or philosophical science? *International Journal of Educational Excellence, 1*(2), 101–110.

Panasuk, R. M., & Bolinger-Horton, L. (2012). Integrating history of mathematics into the curriculum: What are the chances and constraints? *International Electronic Journal of Mathematics Education, 7*(1). Retrieved from www.mathedujournal .com/dosyalar/IJEM_v7n1_1.pdf

Schubring, G., Cousquer, E., Fung, C-I., El Idrissi, A., Gispert, H., Heide, T., Ismael, A., Jahnke, N., Lingard, D., Nobre, S., Philippou, G., Pitombeira de Carvalho, J., & Weeks, C. (2000). History of mathematics for trainee teachers. In J. Fauvel, & J. van Maanen (Eds.), *History in mathematics education: The ICMI study* (pp. 91–142). Dordrecht, The Netherlands: Kluwer Academic.

Selvi, K. (2008). Phenomenological approach in education. In A-T. Tymieniecka (Ed.), *Analecta Husserliana: The yearbook of phenomenological research, Volume 95* (pp. 39–53). Dordrecht, The Netherlands: Springer.

Siu, M-K. (2004). No, I don't use history of mathematics in my class. Why? In S. Kaijser, F. Furinghetti, & A. Vretblad (Eds.), *History and pedagogy of mathematics* (pp. 365–376). Uppsala, Sweden: HPM.

Siu, M-K. (2013). Zhi yì xíng nán (Knowing is easy and doing is difficult) or vice versa? A Chinese mathematician's observation on HPM (history and pedagogy of mathematics) activities. In B. Sriraman, J. Cai, K. Lee, L. Fan, Y. Shimizu, C. S. Lim, & K. Subramiam (Eds.), *The first sourcebook on Asian research in mathematics education: China, Korea, Singapore, Japan, Malaysia and India,* (pp. 1–18). Retrieved from http://hkumath.hku.hk/~mks/18HPM_MKSiu _APAedit_Rev_Feb28_2012(Edited).pdf

Siu, M-K., & Tzanakis, C. (2004). History of mathematics in classroom teaching— Appetizer? Main course? Or dessert? *Mediterranean Journal for Research in Mathematics Education, 3*(1–2), v–x.

Smestad, B. (2009). *Teachers' conceptions of history of mathematics.* Retrieved from http://home.hio.no/~bjorsme/HPM2008paper.pdf

Swetz, F. J. (1994). *Learning activities from the history of mathematics.* Portland, ME: Walch Publishing.

Tzanakis, C., & Arcavi, A. (2000). Integrating history of mathematics in the classroom: An analytic survey. In J. Fauvel, & J. van Maanen, (Eds.), *History in mathematics education* (pp. 201–240). Dordrecht, The Netherlands: Kluwer Academic.

Chapter 8

Professional Development to Support the Learning and Teaching of Geometry

Examining the Impact on Teacher Knowledge, Instructional Practice, and Student Learning in Two Contexts

Jennifer K. Jacobs, Karen Koellner, Nanette Seago,
Helen Garnier, and Chao Wang[1]

INTRODUCTION

Professional development (PD) that sets ambitious goals for teacher learning leading to instructional change is in high demand. Supporting teachers to skillfully implement the common core state standards is currently a key focus of many mathematics PD efforts, with an eye toward helping teachers gain mathematical knowledge for teaching and fluency in instructional decision-making that supports students' learning of challenging content. The learning and teaching geometry (LTG) intervention consists of well-specified PD materials that engage teachers in learning complex geometric concepts targeted in the standards through viewing and discussing videocases.

This chapter describes a group-randomized study to examine the effectiveness of LTG intervention in two locations within the United States. The intervention was found to have a positive effect on teachers' knowledge and instructional quality and to promote student knowledge gains in a targeted content area. Additionally, there were differences across the two locations in terms of the nature and extent of the LTG PD's impact. The study reported in this chapter explores the gains teachers made along with possible explanations for the differences between the locations and implications for future research.

THEORETICAL FRAMEWORK AND OBJECTIVES

The mathematics education literature indicates consistent and positive links between teachers' mathematical knowledge, instructional quality, and student outcomes (e.g., Desimone, 2009; Jacob, Hill, & Corey, 2017). A commonly accepted theoretical trajectory of teacher learning suggests the following pathway: gains in teacher knowledge lead to subsequent changes in instruction that, in turn, support improved student learning (Desimone, 2009; Yoon, Duncan, Lee, Scarloss, & Shapley, 2007). In other words, the impact of increased teacher knowledge on student knowledge growth is thought to be mediated by improved instructional quality (Kersting, Givvin, Sotelo, & Stigler, 2010). There is some preliminary evidence that this hypothesized trajectory holds true based on studies measuring all three elements (Baumert et al., 2010; Kersting, Givvin, Thompson, Santagata, & Stigler, 2012).

A central challenge for the field of teacher PD is how to design interventions that target teacher knowledge while also maintaining a focus on instructional practice and student learning. A number of researchers have worked to address this challenge and there is now a strong research base delineating critical design aspects of effective PD (e.g., Borko, Jacobs, & Koellner, 2010; Desimone, Porter, Garet, Yoon, & Birman, 2002; Ingvarson, Meiers, & Beavis, 2005; Penuel, Fishman, Yamaguchi, & Gallagher, 2007). At the same time, studies of PD outcomes yield a mixed picture. Although some PD programs that adhere to design recommendations by the literature have produced encouraging results (e.g., Franke, Carpenter, Levi, & Fennema, 2001; Kutaka et al., 2017; Taylor, Roth, Wilson, Stuhlsatz, & Tipton, 2017), others have proven much less successful (e.g., Jacob et al., 2017; Santagata, Kersting, Givvin, & Stigler, 2010).

The Value of Video-Based PD

Designing PD that incorporates classroom video has become increasingly popular as numerous studies have documented the potential of viewing and discussing video to foster teacher learning. Two recent, comprehensive literature reviews highlight the substantial promise of video-based PD (Gaudin & Chaliès, 2015; Major & Watson, 2018). Gaudin and Chaliès reviewed 255 articles related to video viewing in teacher education and PD, concluding that classroom video is an especially productive learning tool with regard to improving instructional practice. They argue, "The value of video use by teachers lies principally in the opportunity to raise teachers' quality of instruction" (p. 59).

Major and Watson (2018) limited their review to the use of video in PD programs serving in-service teachers. Based on 82 studies, they concluded that not only is video a powerful and effective tool to promote improvements in classroom practice, it has the potential to enhance teacher cognition as well.

When purposefully selected and integrated into a PD framework, video footage from classroom lessons can advance specific learning goals such as the improvement of particular instructional practices (Blomberg, Sherin, Borko, & Seidel, 2013). Targeted and scaffolded video viewing engages teachers in reflection, providing an opportunity to "mentally simulate instructional action" (Blomberg, Sherin, Renkl, Glogger, & Seidel, 2014, p. 445). Engaging in this type of self-reflective process while viewing video contributes to the development of teachers' professional vision, a type of knowledge that includes the ability to notice and make sense of events in an instructional setting (Sherin, Jacob, & Philipp, 2011; Sherin & van Es, 2009).

Van Es and Sherin (2010) found that as teachers became more adept at analyzing videos through the lens of student thinking, they carried this focus into their classrooms in order to elicit, probe, and build on students' ideas. Similarly, research by Kersting and colleagues (Kersting, 2008; Kersting et al., 2012) suggests that teachers' ability to analyze and interpret video clips of mathematics classrooms is positively related to the instructional quality of their own mathematics lessons.

Video Promotes Community Formation and Collaborative Inquiry

Within a PD context, videos of classrooms can be understood as boundary objects (Kazemi & Hubbard, 2008) in the sense that videos help to maintain coherence across multiple and intersecting social worlds (Star & Griesemer, 1989). As boundary objects, videos enable a shared space within which teachers and the PD facilitator can form a professional community. Videos serve as analytical sites that enable multiple viewpoints and ideologies to surface from multiple parties (Miller & Zhou, 2007).

Furthermore, teachers have a dual role in PD settings: they are both teacher and student, or learner. The analysis of video can support teachers to take up these roles in specified moments and traverse boundaries as needed. Video can encourage "perspective taking" (Akkerman & Bakker, 2011) by helping teachers see themselves in both roles and reflexively understand the needs and limitations of these two identities.

Although viewing video is in many ways a self-directed exercise in which teachers attend to topics of their own interest and construct personally relevant knowledge, when used as a boundary object in a PD setting, video viewing can encourage productive conversations and community building (Borko et al., 2008; van Es & Sherin, 2008). Using video to support collaborative inquiry and engagement can anchor dialogue and lead to collective ideas for improvement (Koh, 2015). For example, van Es (2012) reported that in video club PD meetings, over time the teachers held increasingly substantive discussions about problems of practice that were pertinent to the group, and the community adopted more collegial discourse norms.

Situative Theory and the Intersection of Video, Community, and Learning

Situative theory suggests that learning is situated in a particular context, socially organized, and distributed across individuals, artifacts, and tools (Putnam & Borko, 2000). Learning emerges through authentic activities and social interactions within a community (Brown, Collins, & Duguid, 1989; Lave & Wenger, 1991). The learning setting plays a critical role, and each learning setting is unique with regard to the types of knowledge and experiences that the participants bring with them (Bransford, Brown, & Cocking, 2000).

Using classroom video in PD situates teacher learning firmly within their everyday routines of practice. In addition, the task of analyzing and making sense of video presents a high degree of complexity, providing teachers with motivation to socially engage and interact cooperatively (Blomberg et al., 2013). Hatch, Shuttleworth, Jaffee, and Marri (2016) propose three elements that shape teachers' learning from video in a professional development setting: (1) the properties or affordances of the videos, (2) the background knowledge and experiences that teachers bring to the PD, and (3) the social factors and activities in which videos are viewed and discussed by the participating teachers.

Hatch et al. (2016) argue, "The situated perspective highlights that the affordances for learning from video emerge from the interaction of all three" (p. 276). In other words, situative theory helps make clear that teacher learning from a given video-based PD occurs as a result of the relationship between selected video representations of teaching, participants' unique background, and socially organized viewing experiences.

How Might Context Influence What Teachers Do and Learn within PD?

A situative perspective suggests that groups of teachers who take part in PD workshops using the same materials and with the same facilitator but situated within different educational contexts (e.g., different geographical locations within the United States) might have very different learning opportunities and experiences. Even within a highly specified, video-based PD in which resources and facilitation materials are provided to ensure a particular experience (Koellner & Jacobs, 2015), the nature of teachers' conversations and the degree to which their learning is impacted is unlikely to be constant across contexts. For example, precisely what teachers notice about a video clip can yield unique perspectives, conversations, and reflections within a given group of participants.

The knowledge and prior experiences teachers bring with them to the workshop is also likely to be quite different across locations. Additionally, as noted previously, during PD workshops the participating teachers develop a unique professional community that serves as the social unit within which learning takes place. As Webster-Wright (2009) argued, "learning always occurs in a context, as has been highlighted by the range of research into the social, situated nature of learning" (p. 723).

At present very little is known about the degree to which context impacts teachers' learning from PD. Do teachers from different locations who participate in the same intervention with the same facilitator experience differential impacts on their knowledge, practice, and student learning?

We argue there are at least three reasons to conjecture that the answer is likely to be yes: (1) wide variation exists across geographic locations (i.e., the U.S. states) on nearly all educational measures such as teacher-pupil ratios, teacher salaries, teacher experience levels, teacher turnover, student achievement, and PD opportunities (Grissmer, Flanagan, Kawata, Williamson, & LaTourrette, 2000; Wei, Darling-Hammond, & Adamson, 2010); (2) schools and districts have distinct organizational and workplace climates that affect teachers' formal and informal learning and growth opportunities (Avalos, 2011; Cobb, McClain, Lamberg, & Dean, 2003; Firestone, Mangin, Martinez, & Polovsky, 2005); and (3) teachers form unique professional learning communities within PD contexts that impact their collective learning and individual growth (Grossman, Wineburg, & Woolworth, 2001; Little, 2003).

Borko (2004) argues for the importance of research on well-specified PD programs implemented in multiple sites with multiple facilitators in order for the field to move beyond "existence proofs" of effective PD. In our study, we implemented a video-based mathematics PD program in different contexts—specifically, two unique geographic locations within the United States. Both programs were led by the same individual using highly detailed and extensive facilitation materials to ensure consistency in implementation. As such, our research joins a very limited body of literature that is able to compare the effectiveness of a given PD program across multiple contexts.

Yet even widely scaled professional learning endeavors such as the National Writing Project (Academy for Educational Development, 2002; Gallagher, Woodworth, & Arshan, 2015) tend not to disaggregate their data across sites to examine location effects, particularly in a quantitative fashion. Even when research on PD is conducted and reported across sites, variation in effectiveness is generally attributed to differences in facilitation or other aspects of the implementation of the PD (e.g., Bell, Wilson, Higgins, & Mc-Coach, 2010).

The Need for Professional Learning in the Area of Geometric Transformations

A critical feature of effective PD is that it addresses a problem of practice, meaning that it meets the professional needs of teachers (Scribner, 1999). Starting in middle school, the common core state standards for mathematics (CCSSM) contain a strong and consistent focus on geometric transformations—including their mathematical properties, how they can be sequenced, and their effect on two-dimensional figures in a coordinate plane.

For example, seventh graders are expected to understand how to scale drawings of geometric figures. The eighth-grade standards encourage an in-depth understanding and application of geometric transformations along with connections between proportional reasoning, slope, and linearity to geometric transformations and similarity. The high school standards continue to promote a transformations-based approach to understanding congruence, similarity, and the proof of various relationships among geometric figures based on similarity. This increased emphasis on transformational geometry represents a major difference from previous state standards (Teuscher, Reys, & Tran, 2016; Tran, Reys, Teuscher, Dingman, & Kasmer, 2016).

As Teuscher and colleagues (2016) explain, "most high school students are taught that for two figures to be similar they must have corresponding congruent angles and corresponding side lengths proportional. This again brings up the question, why? With an understanding of geometric transformations, students can answer the why question as they make connections between these concepts" (p. 11). Several reports have called on professional developers to create materials that emphasize transformations-based geometry in accordance with the CCSSM (McCallum, 2011; Sztajn, Marrongelle, Smith, & Melton, 2011). PD materials that help support teachers in this area can play a timely role in the present mathematics education environment.

Teachers generally agree that they need much more support to learn and effectively implement the common core state standards in their classrooms. In an online survey conducted by the EPE Research Center, a majority of teachers responded that they had participated in some PD related to the CCSS and agreed that the standards would help them improve their classroom practice

(Gewertz, 2013). However, across several studies teachers reported only a moderate level of preparedness to teach the standards to their students, and they indicated a strong desire for professional learning opportunities to ensure alignment between their curricular materials and the standards (Gewertz, 2013; Authors, 2015b; Roth McDuffie et al., 2017).

The Learning and Teaching Geometry PD Intervention and Field Test

The LTG program (developed with funding from National Science Foundation award #0732757) is a video-based mathematics professional development intervention targeted to teachers serving grades 6–12. The intervention consists of 54 hours of PD focused on improving teaching and learning of mathematical similarity based on geometric transformations (Seago et al., 2017). The program is designed to be implemented by a knowledgeable facilitator using a set of provided resources to engage teachers in a specified learning trajectory aligned with multiple middle and high school CCSSM (Seago et al., 2013).

The intervention includes a 30-hour foundation module followed by four 6-hour extension modules that explore related topics such as using appropriate representations and tools and supporting English language learners. The LTG PD intervention targets teachers' mathematics knowledge for teaching geometric similarity and is intended to inform both their content and pedagogical content knowledge in ways that foster student learning of this topic.

The foundation module was field tested in eight sites throughout the United States (Borko, Jacobs, Seago, & Mangram, 2014; Seago, Jacobs, Heck, Nelson, & Malzahn, 2014). In total, the pilots involved 127 participants: 87 treatment teachers and 40 comparison teachers. The treatment and comparison teachers completed a series of pre- and post-knowledge assessments. The knowledge assessments included a multiple-choice content assessment and two embedded assessments (i.e., incorporated in the PD workshops). Two of the field test sites agreed to administer a student assessment to the students of both the treatment ($n = 162$) and comparison teachers ($n = 104$). The student assessment was a multiple-choice test that focused on geometric similarity.

Treatment teachers demonstrated significant knowledge gains relative to the comparison teachers on all of the assessments. On the student assessment, students of the treatment teachers on average demonstrated significantly higher gains compared to students of the comparison teachers. Taken together, the results of the field study provide evidence of the promise of the foundation module for achieving the intended teacher and student knowledge outcomes.

However, the finalized version of the foundation module was not used in the field test nor were any of the extension modules. Other limitations included the fact that the samples were based on convenience without random assignment; the pilot groups used different versions of the materials (depending on which version was available at the time) with different facilitators and different teacher populations; and the data collected were relatively limited (i.e., no use of knowledge assessments farther removed from the PD content and no data on classroom practices).

METHODOLOGY

In this study, we present findings from an LTG efficacy study (NSF award #1503399) that was designed to explore the effectiveness of full LTG PD intervention using a group-randomized experimental design. We consider the overall effectiveness of PD on critical aspects of teacher learning and explore geographic, location-based differences in teacher and student learning.

Research Questions

Our research questions were as follows:

1. Did taking part in the LTG PD impact teacher and student learning?
   a. Did teachers who participated in the LTG PD increase their knowledge of geometry and transformations-based geometry beyond that of teachers in the control group?
   b. Did teachers who participated in the LTG PD improve the quality of their classroom instruction beyond that of teachers in the control group?
   c. Did students taught by teachers who participated in the LTG PD increase their knowledge beyond that of students taught by teachers in the control group?
2. Did the impact of the LTG PD vary across the two locations?

Method

This study used a group-randomized design to examine the effectiveness of LTG intervention. The intervention consisted of well-specified PD materials concerned with teachers learning complex geometric concepts.

Sample

Participants in the LTG efficacy study were 103 secondary mathematics teachers serving grades 6–12 (47% middle school, 53% high school) from two geographic locations (49% in Location A, 51% in Location B). One of the locations included teachers from multiple school districts and the other location included teachers from a single large district. Both locations included teachers from a range of grade levels and schools who used a variety of curricular resources in their mathematics classrooms.

The selected locations were included in the study for several reasons. First, the two locations exhibited unique student populations with a large degree of diversity both within and across geographic regions of the United States. Location A included teachers from multiple suburban districts surrounding a city in the midwestern United States. These districts had a relatively large Hispanic and English language learner population. Location B included teachers from a single, urban school district in the northeastern region of the United States. The student population in this district was largely non-White, with the largest proportion of students identifying as Black.

A second reason these locations were selected was that district stakeholders in both locations had strong existing collaborative ties with members of the research team as well as an expressed interest in incorporating the LTG PD as a means of improving their teachers' knowledge and instruction in transformations-based geometry. The assistance of these stakeholders was critical to ensuring a large enough sample of teachers could be recruited and that the districts would support two years of ongoing PD workshops. Finally, including two geographically distinct locations enabled a large enough sample size to compare the treatment teachers to a comparison group and to look at the impact of PD within location.

Randomization was conducted at the school level with 32 schools and 49 teachers assigned to the treatment group and 35 schools and 54 teachers assigned to the comparison group. Table 8.1 provides background information on the sample. No statistically significant differences were found by treatment and control groups for the full sample or within the two locations. Three significant differences were detected across locations: teachers in Location A had more years of experience teaching mathematics ($p < 0.05$) whereas more teachers in Location B had graduate degrees ($p < 0.001$) and were certified in special education ($p < 0.01$).

Not all teachers administered both the pre- and post-student assessment for this study, resulting in a reduced sample available for analyses of student knowledge outcomes. The sample with student pre-test and post-test data included 56 teachers and 758 students. One teacher background difference was found with the reduced sample reporting significantly ($p < 0.05$) more

Table 8.1. Teacher Background Survey Responses by Group and Location—Percentage of Teachers

Background	All Teachers (N = 103)	Location A (n = 50)	Location B (n = 53)
Current Grade Level			
Middle school	48	54	40
High school	55	46	60
Grade Levels Taught			
Elementary	21	30	14
Middle school	64	64	64
High school	80	84	76
Years Teaching			
0–5	40	32	48
6–10	28	22	35
11–20	25	36	13
> 20	7	10	4
Years Teaching Math			
0–5	45	36	53
6–10	27	24	30
11–20	22	32	13
> 20	6	8	4
Teaching Certification			
Elementary education	19	26	13
Elementary math	2	4	0
Middle math	45	42	47
Secondary math	80	82	77
Undergraduate or graduate degree in math, math Education	79	64	93

Note. Percentages may be greater than 100 because some participants responded in more than one category.

years of teaching (10 years) compared to the full sample (seven years). Years teaching was used as a covariate in subsequent analyses.

Professional Development Workshops

The LTG PD workshops for treatment teachers in both locations began in summer 2016 and continued through the 2016–2017 academic year. Nine full days of PD were offered to teachers in each location beginning with a five-day foundation module in the summer followed by four days of extension modules during the school year.

No further PD was provided after spring 2017. On average, treatment teachers attended seven days of PD (7.6 for Location A teachers and 6.5 for Location B teachers). In Location A all workshops were held during the school day (with substitutes paid for by the project). Location B teachers likely had a lower attendance rate due to the fact that their academic year workshops were held on weekends per district policy.

Control teachers were offered the opportunity to participate in the same LTG PD workshops (including the foundation module and extension modules) during the 2017–2018 school year once pre- and post-data collection was completed. The same facilitator led all of the workshops in both locations after taking part in an extensive facilitator preparation process that included a multifaceted assessment of fidelity.

Based on this fidelity assessment process, the facilitator was deemed highly capable of using the PD materials as intended and making appropriate, context-based decisions and adaptations (Jacobs, Seago, & Koellner, 2017). In all cases, the workshops were held first in Location A and then in Location B for pragmatic reasons, largely due to the timing of the academic years in the two locations.

Measures

Pre- and post-data collected on the effectiveness of the LTG PD included measures of teacher and student knowledge, videos of classroom instruction, and anonymous teacher reflections on the professional development. Baseline data for all teachers was collected in spring 2016 (prior to the summer workshops for the treatment teachers), and post-data was collected in spring 2017.

Teacher Knowledge

Four assessments were used to examine impact on the teachers' knowledge: one developed by the Center for Research in Mathematics and Science Teacher Development, University of Louisville, and three created by Horizon Research. The assessment developed by the University of Louisville is part of a series called Diagnostic Teacher Assessments in Mathematics and Science (DTAMS; Saderholm, Ronau, Brown, & Collins, 2010). For this study we administered the DTAMS geometry/measurement assessment, which serves as a general assessment of teachers' geometry and measurement content and pedagogical content knowledge. The DTAMS assessment is divided into two domains: "knowledge" and "mathematics subject."

Items in the knowledge domain include facts, conceptual understanding, reasoning/problem solving, and pedagogical content knowledge. Items in the mathematics subject domain include 2-D geometry, 3-D geometry, transformational/coordinate geometry, and measurement. Scoring of the DTAMS assessment, which comprises both closed and open-ended items, was conducted by the DTAMS development team in accordance with their usage policy.

Horizon Research developed three pre-post assessments to measure the impact of the LTG PD on teachers' knowledge of geometric similarity: a multiple-choice assessment and two embedded assessments (Seago et al., 2014). These were the same measures used in the field study described earlier. They were developed in tandem with the creation of the LTG PD materials and are the most proximal measures of teacher learning from the intervention. The Horizon multiple-choice assessment covered five content areas: dilation, properties of similarity, ratio and proportion, scaling, and congruence transformations (i.e., translation, rotation, and reflection).

The items were compiled and modified from existing state, national, and international assessment sources. About half of the items were purely content based; the rest were set in contexts that situated them in the work of teaching geometry. The items were validated by the LTG developers as accurate and appropriate to the content emphasis of the intervention.

Horizon Research's embedded assessments consisted of a mathematics task and a videocase analysis task. These tasks existed within the foundation module and could be administered either as part of the LTG PD (i.e., early in the PD and again later in the PD) or separately (i.e., to the comparison teachers). The mathematics task included three open-ended items that focused primarily on the properties of similarity.

The videocase analysis task also included three open-ended items that focused primarily on dilation. Horizon developed initial scoring rubrics for the embedded assessments. A modified version of the rubrics was applied by our research team after achieving initial inter-rater agreement of 80–93% on the items within the mathematics task and 85–97% on the items within the videocase analysis task.

Classroom Observations

Teachers' mathematics lessons were videotaped and rated using the math in common teacher observation protocol (Perry, Seago, Burr, Broek, & Finkelstein, 2015). This protocol, developed as part of WestEd's math in common study of K–12 mathematics instruction, incorporates eight items that capture various elements of lesson quality. The protocol is adapted from the mathematical quality of instruction instrument (Hill, 2014) and teaching for robust understanding (Schoenfeld, 2013; Schoenfeld, Floden, & the Algebra Teaching Study and Mathematics Assessment Project, 2014) coding instruments. Three broad constructs (richness of the mathematics, engagement in practices, and mathematics content) were created by generating average scores from selected items on these instruments, as described below.

Items adapted from the mathematics quality of instruction instrument addressed two broad constructs: richness of the mathematics and student engagement in practices recommended by the common core state standards. Each item was rated on a 4-point scale (1 = not present, 2 = low, 3 = middle, 4 = high) and applied at the lesson level. The items were (1) linking between representations (richness), (2) multiple procedures or solution methods (richness), (3) mathematical sense making (richness), (4) student explanations (engagement), and (5) mathematical reasoning (engagement).

Items adapted from the teaching for robust understanding instrument focused on a single construct: the mathematical content in the lesson. Each item was rated on a 3-point scale (1 = novice, 2 = apprentice, 3 = expert) and applied at the lesson level. The items were (1) attention to accurate and well-justified mathematics, (2) access to mathematical content, and (3) agency, authority, and identity.

Two members of the research team established initial inter-rater agreement of 92.5% on the math in common protocol overall and 80–100% on each item. To ensure that they were applying the items consistently throughout the coding process, the coders established midpoint reliability approximately halfway through the set of videotaped lessons. Midpoint inter-rater reliability was 91% on the protocol overall and 80–100% on each item.

Student Knowledge

In addition to the teacher knowledge assessments described earlier, Horizon Research developed an assessment to measure the impact of the LTG PD intervention on student knowledge. This multiple-choice assessment was constructed based on cognitive interviews with students to ensure the items were clear, plausible, and had content validity. The assessment targeted students' knowledge of transformations-based geometry in five content areas: dilation, properties of similarity, ratio and proportion, scaling, and congruence transformations. These were the same content areas targeted by the Horizon multiple-choice teacher knowledge assessment.

Teacher Reflections

Treatment teachers provided daily written reflections by responding to open-ended prompts in an anonymous online survey at the end of each workshop. Although the prompts varied slightly from day to day, teachers typically reported on what they learned, what they struggled with, what they hoped to take back to their classrooms, and what feedback they had for the facilitator. Their responses were coded by two members of the research team for tone (positive, negative, in-between) and the nature of their self-assessed learning (content, pedagogy, both, neither). Inter-rater agreement of at least 80% was established for both codes.

ANALYSIS

The primary research question focused on estimating the effects of the LTG PD workshops on teacher and student learning. The secondary research question explored the possible impact of different contexts—in this case, geographic locations, on the effectiveness of the LTG PD. To investigate teacher learning, initial analyses of variance were used to identify any differences between the treatment and control groups on teacher background or baseline knowledge and instruction measures.

Analyses of variance also were used to identify the impact of the intervention on gains in teacher knowledge and improvement in quality of instruction. The same analyses were repeated within each location to help answer the second research question.

To explore the effect of the LTG PD on students' learning of geometry, we used hierarchical linear models (HLM 7.0; Raudenbush, Bryk, Cheong, Congdon, & Du Toit, 2011) due to the multilevel structure of the data with students nested within teachers. Although a three-level hierarchical model accounts for the nested data structure of students in classrooms with the same teacher and teachers within schools, the sparse number of teachers per school (56 teachers in 41 schools) made it difficult to reliably estimate the variance between teachers within schools. Therefore, we limited analyses to a two-level model with a student level and a classroom level. The classroom-level predictor was the effect of the intervention on student geometry knowledge gains. Two teacher background covariates were included: years teaching in general and current grade level teaching. Level 1 included 758 students; Level 2 included 56 teachers. The same model of student knowledge gains was used to estimate the effect of the LTG PD within Locations A and B.

RESULTS

The LTG intervention was found to have a positive effect on teachers' knowledge and instructional quality and to promote student knowledge gains in a targeted content area. Additionally, there were differences across the two locations in terms of the nature and extent of the LTG PD's impact. Three categories of impact were identified: impact on teacher knowledge, impact on classroom practice, and impact on student knowledge. Within each category two types of changes were denoted: impact on teacher knowledge focused on overall changes in knowledge, and changes in teacher knowledge by context.

Overall Changes in Knowledge: Treatment versus Comparison Teachers

The treatment teachers made significant gains over time in the knowledge areas shown in Table 8.2. No pre-test differences were found, but the treatment group's average post-test scores were significantly ($p < .01$) larger than the control group's in congruence transformations, dilations, similarities, and total geometry. Across all four teacher knowledge assessments, a significant effect of the LTG intervention was found for 16% of the content subscales (including total scores). The average gains for the treatment group were significantly larger than for the control group in dilations ($p < .01$, ES = .54), similarity ($p < .05$, ES = .49), and total geometry for the embedded math task ($p < .01$, ES = .60).

Taken together, the results indicate that teachers who participated in the LTG PD increased their knowledge in several of the focal content areas. However, on the most distal assessment of geometry knowledge—the DTAMS geometry/measurement assessment—no differences were found between the treatment and comparison teachers.

The treatment group mainly showed improvement on the highly proximal embedded assessments. Thus the findings are not particularly robust across measures, suggesting that (1) the teachers' knowledge gains may not be generalizable across mathematical contexts, (2) other teacher knowledge assessments might be necessary to detect meaningful gains, or (3) the LTG intervention might not have had a particularly strong impact on teachers' knowledge.

Changes in Teacher Knowledge by Context: Patterns within Locations A and B

Table 8.3 shows changes in teachers' knowledge within each of the two locations. In Location A, the treatment group made significant gains in measurement, dilations, and total geometry, and those gains were significantly larger than the control group's for dilations ($p < .01$, ES = .78) and total geometry ($p < .01$, ES = .77). Treatment teachers in Location B similarly made gains in dilations but they also made gains in different content areas: reasoning, problem solving, and congruence transformations.

None of the teacher knowledge gains in Location B were significantly larger than control gains, although the moderate effect size for congruence transformations ($p = .08$, ES = .53) is worth noting in light of the reduced sample size. No pre-test or post-test group differences were found within location.

Table 8.2. Means and Standard Deviations of Average Teacher Scores on Geometry Measures over Time (N = 103 Teachers)

Content	Treatment (n = 49)		Control (n = 54)		Change over Time		
	Pre-PD M (SD)	Post-PD M (SD)	Pre-PD M (SD)	Post-PD M (SD)	Treatment M	Control M	T-C Difference M
Reasoning, problem solving (DTAMS)	6.52 (2.25)	7.25 (2.43)	5.98 (2.27)	6.71 (2.53)	.73**	.73**	.00
Measurement (DTAMS)	5.38 (2.00)	6.40 (1.95)	5.20 (2.17)	5.96 (2.14)	1.02**	.76*	.26
Congruence transformations (Horizon multiple choice)	3.04 (1.03)	3.43 (.91)	2.74 (1.13)	2.81 (1.06)	.39**	.06	.33
Dilations (Horizon embedded videocase)	14.13 (5.10)	17.04 (3.04)	13.26 (6.04)	13.13 (6.03)	2.91***	-.13	3.04**
Properties of similarity (Horizon embedded videocase)	3.09 (1.74)	3.65 (1.89)	2.90 (2.00)	2.58 (2.02)	.56*	-.33	.89*
Total geometry (Horizon embedded math task)	9.02 (2.86)	10.47 (3.34)	8.31 (3.01)	7.48 (3.23)	1.45**	-.83	2.28**

*$p < .05$; **$p < .01$; ***$p < .001$.

Table 8.3. Means and Standard Deviations of Average Teacher Scores on Geometry Measures over Time within Location (N = 103 Teachers)

Content	Treatment		Control		Change over Time		
	Pre-PD M (SD)	Post-PD M (SD)	Pre-PD M (SD)	Post-PD M (SD)	Treatment M	Control M	T-C Difference M
Location A (n = 50; 25 treatment, 25 control)							
Reasoning, problem solving (DTAMS)	6.56 (1.96)	7.12 (2.30)	6.28 (2.46)	7.32 (2.29)	.56	1.04**	−.48
Measurement (DTAMS)	5.52 (2.06)	6.56 (1.81)	5.72 (2.13)	6.60 (1.66)	1.04*	.88	.16
Congruence transformations (Horizon multiple choice)	3.42 (.78)	3.63 (.71)	2.88 (1.15)	3.00 (1.18)	.21	.13	.08
Dilations (Horizon embedded videocase)	15.17 (4.53)	17.96 (1.81)	15.28 (4.86)	14.36 (5.14)	2.79**	−.92	3.71**
Properties of similarity (Horizon embedded videocase)	3.50 (1.56)	4.08 (1.82)	3.40 (1.76)	3.12 (1.88)	.58	−.28	.86
Total geometry (Horizon embedded math task)	9.40 (2.93)	11.72 (1.90)	7.96 (3.26)	8.04 (2.48)	2.32***	.08	2.24**
Location B (n = 53; 24 treatment, 29 control)							
Reasoning, problem solving (DTAMS)	6.48 (2.57)	7.39 (2.61)	5.67 (2.06)	6.08 (2.65)	.91*	.42	.49
Measurement (DTAMS)	5.22 (1.95)	6.22 (2.13)	4.67 (2.12)	5.29 (2.40)	1.00	.63	.37
Congruence transformations (Horizon multiple choice)	2.64 (1.14)	3.23 (1.07)	2.61 (1.12)	2.61 (.89)	.59**	.00	.59
Dilation (Horizon embedded videocase)	13.00 (5.54)	16.05 (3.77)	11.19 (6.49)	12.00 (6.65)	3.05*	.59	2.46
Properties of similarity (Horizon embedded videocase)	2.64 (1.84)	3.18 (1.89)	2.44 (2.14)	2.07 (2.06)	.55	−.37	.92
Total geometry (Horizon embedded math task)	8.63 (2.80)	9.17 (4.00)	8.62 (2.80)	7.00 (3.74)	.54	−1.62*	2.16

*p < .05; **p < .01; ***p < .001.

Similar to the findings for the whole sample, teacher knowledge impact within each of the two locations appeared limited in scope and generalizability, although groups in both locations gained knowledge about dilation. In other content areas, teachers differed in the nature of their learning despite the fact that both groups were led by the same facilitator who adhered to highly specified workshop agendas.

Impact on Classroom Practice

Impact on classroom practice focused on overall changes in instruction and changes in instruction by context.

Overall Changes in Instruction: Treatment versus Comparison Teachers

Compared to the impact on teacher knowledge, impact of the LTG PD on classroom practice was considerably more robust and pervasive. Table 8.4 presents teachers' average ratings and gains on the three instructional practice constructs: richness of the mathematics, engaging students in mathematical practices recommended by the common core state standards, and mathematics content.

The control group was rated significantly higher on the pre-test in all three constructs but no differences were found on the post-test, meaning that the LTG PD appears to have supported the treatment group in catching up to their control peers. The treatment group significantly improved the quality of their instruction beyond the control group in both engaging students ($p < .01$, ES $= .61$) and mathematics content ($p < .01$, ES $= .60$).

Table 8.4. Means and Standard Deviations of Average Teacher Ratings on Observed Quality of Instruction over Time (N = 88 Teachers)

	Treatment (n = 44)		Control (n = 44)		Change over Time		
	Pre-PD	Post-PD	Pre-PD	Post-PD	Treatment	Control	T-C Difference
Instruction	M (SD)	M (SD)	M (SD)	M (SD)	M	M	M
Richness of mathematics	1.96 (.73)	2.28 (.89)	2.31 (.85)	2.41 (.92)	.33**	.10	.23
Student engagement in mathematical practices	1.82 (.76)	2.18 (.96)	2.39 (1.04)	2.27 (.95)	.37**	−.12	.49**
Mathematics (content, access, agency)	1.87 (.53)	2.05 (.65)	2.23 (.57)	2.03 (.61)	.18	−.19*	.37**

*$p < .05$; **$p < .01$.

Changes in Instruction by Context: Patterns within Locations A and B

Some location differences emerged in the degree and nature of instructional quality improvement (see Table 8.5). The control group's higher pre-test ratings seen in Location A for engaging students ($p < .05$) and in Location B for richness of mathematics ($p < .05$) and the mathematics content ($p < .05$) disappeared on the post-test given that the control group's instructional quality largely remained stagnant or declined. The treatment groups in both locations shared significant improvement beyond the control group in engaging students (Location A: $p < .05$, ES = .61; Location B: $p < .05$, ES = .60), and the treatment group in Location B also improved in mathematics content ($p < .05$, ES = .73).

Table 8.5. Means and Standard Deviations of Average Teacher Ratings on Observed Quality of Instruction over Time within Location (N = 88 Teachers)

	Treatment		Control		Change over Time		
	Pre-PD	Post-PD	Pre-PD	Post-PD	Treatment	Control	T-C Difference
Instruction	M (SD)	M (SD)	M (SD)	M (SD)	M	M	M
	Location A (n = 47; 23 treatment, 24 control)						
Richness of	2.09	2.35	2.18	2.07			
mathematics	(.73)	(.93)	(.87)	(.87)	.26	−.11	.37
Student engagement							
in mathematical	1.70	2.04	2.25	2.10			
practices	(.67)	(.98)	(1.07)	(.98)	.35*	−.14	.49*
Mathematics							
(content, access,	1.93	2.00	2.22	1.93			
agency)	(.56)	(.74)	(.58)	(.69)	.07	−.29*	.36
	Location B (n = 41; 21 treatment, 20 control)						
Richness of	1.81	2.21	2.47	2.82			
mathematics	(.72)	(.85)	(.83)	(.83)	.40**	.36	.04
Student engagement							
in mathematical	1.95	2.35	2.55	2.47			
practices	(.84)	(.98)	(1.01)	(.89)	.40*	−.08	.48*
	1.81	2.10	2.23	2.15			
Mathematics	(.50)	(.56)	(.56)	(.47)	.29**	−.08	.37*

*$p < .05$; **$p < .01$.

Impact on Student Knowledge

Impact on student knowledge focused on overall changes in knowledge and changes in student knowledge by context.

Overall Changes in Knowledge: Treatment versus Comparison Students

Table 8.6 shows the average pre- and post-test scores and gains students made within each teacher group. Pre-existing differences appeared on the dilations and ratio pre-test with significantly ($p < .05$) higher averages for the treatment group but no differences on the post-test. Both treatment and control groups of students similarly made gains in all geometry content areas except proportion, which showed no change. Analyses of student gains identified one effect of the LTG PD on student knowledge of congruence transformations with a significantly larger average gain for the treatment group beyond the control group ($p = .048$, ES = .23).

Multi-level comparisons were used to estimate treatment and control group differences and the impact of the LTG PD on students' learning of geometry. Preliminary analysis of the initial unconditional HLM model confirmed that for each content area there was significant ($p < .001$) variation among classrooms available to be explained by the model. Additional preliminary analyses confirmed that neither of the teacher covariates significantly contributed to the model and both were omitted in favor of a more parsimonious model. The LTG PD explained 8.5% of the variance in student congruence transformations gain.

Table 8.6.   Means and Standard Deviations of Student Gains on Geometry Knowledge over Time (N = 758 Students)

Content	Treatment (n = 375)		Control (n = 383)		Change over Time		
	Pre- PD	Post- PD	Pre- PD	Post- PD	Treatment	Control	T-C Difference
	M (SD)	M (SD)	M (SD)	M (SD)	M	M	M
Dilations	1.54 (.95)	1.73 (.94)	1.27 (.86)	1.42 (.91)	.19***	.16**	.03
Scaling	2.81 (1.56)	3.30 (1.65)	2.29 (1.44)	2.76 (1.58)	.51***	.48***	.03
Ratio	2.68 (1.33)	3.03 (1.36)	2.16 (1.35)	2.72 (1.36)	.39***	.57***	−.18
Proportion	1.20 (.74)	1.24 (.70)	1.06 (.72)	1.11 (.69)	.04	.06	−.02
Congruence transformations	1.57 (.99)	2.09 (1.02)	1.53 (.91)	1.80 (1.07)	.51***	.28***	.23*

*p < .05; **p < .01; ***p < .001.

Changes in Student Knowledge by Context: Patterns within
Locations A and B

Student knowledge scores at each location generally reflected the same patterns as the full sample. At both locations the treatment group pre-test averages were significantly ($p < .05$) higher in the area of dilations and no differences were detected at post-test (see Table 8.7). Treatment and control groups in both locations made significant gains in all areas except proportion. In the domain of congruence transformations, both groups in each location made significant gains; however, in Location B the treatment group's gain significantly outdistanced the control group's gain.

Based on the HLM analyses shown in Table 8.8, participation in the LTG PD significantly predicted 49.6% of the variance in student gain in congruence transformations knowledge ($p < .001$, ES = .41) in Location B. The LTG PD did not have a significant effect in Location A. This lack of effect in Location A might be explained, in part, by the treatment and control students making almost identical average gains of .37 to .38. Furthermore, Location A's control group gain was significantly larger than Location B's control group gain ($p < .01$).

The HLM models provide evidence that taking part in the LTG PD, together with using instructional practices recommended by the standards, supported growth in student knowledge in the area of congruence transformations. It is important to note that these student knowledge gains were limited in scope to congruence transformations and did not apply to all of the mathematical domains targeted by the LTG PD.

CONCLUSION

This study explored the gains teachers made along with possible explanations for the differences between the locations. Conclusions are drawn concerning the LTG intervention and the differences across the two locations.

LTG PD Intervention Supports Critical Gains

Our primary research question focuses on the impact of the LTG PD on teachers' knowledge, classroom practices, and student learning. The highly proximal embedded assessment captured significant teacher knowledge gains beyond the control group; however, these gains were restricted to only some of the targeted mathematical domains, including congruence transformations (the rigid transformations: translation, rotation, and reflection), dilation (a non-rigid transformation), and measurement. Although these topics were a

Table 8.7. Means and SDs of Average Student Gains on Geometry Measures over Time by Location (N = 758 Students)

| | Treatment | | Control | | Change over Time | | |
| | Pre-PD | Post-PD | Pre-PD | Post-PD | Treatment | Control | T-C Difference |
Horizon Subject	M (SD)	M (SD)	M (SD)	M (SD)	M	M	M
	Location A (n = 397; 203 treatment, 194 control)						
Total	9.44	10.93	8.61	1020			
	(4.01)	(4.47)	(3.72)	(4.24)	1.49***	1.59***	−.10
Dilations	1.51	1.56	1.19	1.46			
	(.94)	(.92)	(.89)	(.90)	.06	.25**	−.19
Scaling	2.69	3.22	2.37	2.81			
	(1.56)	(1.65)	(1.52)	(1.60)	.56***	.41***	.15
Ratio	2.66	2.98	2.38	2.89			
	(1.32)	(1.34)	(1.33)	(1.33)	.38***	.51***	−.13
Proportion	1.10	1.22	1.16	1.18			
	(.74)	(.72)	(.73)	(.73)	.12*	.03	.09
Congruence transformations	1.63	2.01	1.57	1.95			
	(1.00)	(1.09)	(.96)	(1.16)	.38***	.37***	.01
	Location B (n = 361; 172 treatment, 189 control)						
Total	10.01	11.76	7.70	9.25			
	(3.93)	(4.22)	(.3.57)	(3.98)	1.75***	1.55***	.20
Dilation	1.57	1.92	1.34	1.38			
	(.96)	(.93)	(.82)	(.92)	.34***	.07	.27
Scaling	2.96	3.40	2.20	2.72			
	(1.56)	(1.65)	(1.35)	(1.57)	.44***	.56***	−.12
Ratio	2.71	3.09	1.93	2.55			
	(1.34)	(1.38)	(1.34)	(1.37)	.41***	.64***	−.23
Proportion	1.32	1.26	.96	1.04			
	(.73)	(.69)	(.69)	(.65)	−.06	.09	−.15
Congruence transformations	1.50	2.17	1.48	1.66			
	(.97)	(.92)	(.86)	(.94)	.67***	.18*	.49***

*p < .05; **p < .01; ***p < .001.

central focus of the workshops, it is somewhat disappointing that the gains were not more robust across the assessments and more widespread across the targeted areas.

At the same time, it is encouraging that the treatment teachers made substantial improvement in the quality of their mathematics instruction. Importantly, the teachers were filmed teaching any topic of their choosing. Looking more closely at the treatment teachers' gains on the three instructional quality constructs, we found no significant differences by the mathematical topic of their lessons (categorized as geometry, pre-algebra, pre-calculus, statistics/probability, other).

In other words, the teachers' instructional improvements were not limited to a single content area or to only the geometric content covered by the LTG PD. These findings suggest that a PD program can support mathematics teachers' ability to make widespread changes in practice in the desired direction, across grades and topics. Further, these changes occurred despite the lack of larger gains in knowledge, and despite the fact that the content focus of the PD was relatively narrow in scope.

As we know from the mixed literature on the effectiveness of PD efforts discussed in the introduction, generating improvements in classroom instruction is no small feat and is arguably a more important goal than improving teacher knowledge given its closer relationship to student learning. It is interesting to consider whether an improvement trajectory that results in increased student learning requires an initial increase in teacher knowledge or whether changes in practice are sufficient. We do not contend our study answers this question in any definitive manner, but further investigation into the relationship between teacher knowledge gains, classroom practice improvements, and student learning certainly appears warranted.

It is possible that our findings are linked to the particular nature of the LTG PD program with its heavy focus on collaboratively analyzing classroom video. For example, spending a considerable amount of time during workshops watching and critiquing footage from actual mathematics lessons might trigger especially strong gains in certain classroom practices. This is a topic that could be more thoroughly investigated in future studies by comparing the outcomes of video-based PD and non-video-based PD.

Participating in the LTG PD was directly related to student learning of congruence transformations. Congruence transformations was only one of several targeted content areas, and it is interesting to consider why it might be the area in which students demonstrated the largest knowledge increase. Knowledge of rigid motion transformations is widely relevant across grades and content areas and spans several standards in the middle and high school CCSSM.

Although we did not systematically collect data on teachers' instruction of transformations, we can speculate that this topic may have been more deeply and intentionally covered by the treatment teachers as a result of their workshop experiences. For example, the LTG PD encourages teachers to use a set of nine tasks focused primarily on congruence transformations in their classes regardless of grade level or main content focus. Informally, many teachers shared with our research team that they did incorporate these worksheets.

In addition, after proportion, congruence transformations and dilations were the areas in which students scored most poorly on their pre-tests. Given that an understanding of congruence transformations is generally considered foundational for learning dilation, it is promising that students of the treatment teachers exhibited the most growth in this challenging area of geometry.

Differences Across the Two Locations

Most of the literature on the design and implementation of effective PD cautions that context is a critical factor to keep in mind (e.g., Cohen & Ball, 1999; Loucks-Horsley, Stiles, Mundry, Love, & Hewson, 2003). As discussed in the introduction, teacher learning by design takes place within a given social and policy-based context at the school, district, state, and national level. Additionally, a good deal of learning (including that in the LTG intervention) takes place at the community level among teachers who meet face-to-face at regular intervals over an academic year. Given all of these contextual factors, it is not surprising that learning outcomes might differ across locations.

Our second research question addressed the differential impact of the LTG PD based on the geographic locations of the workshops. Including teachers in two distinct areas of the United States enabled us to explore context-based differences, which did in fact exist. Although the LTG PD positively impacted both Location A and Location B treatment groups in multiple domains of interest, overall the impact appeared to be more robust in Location B.

The treatment teachers in Location A exhibited knowledge gains in more content areas relative to a comparison group than the treatment teachers in Location B; however, treatment teachers in Location B improved their instructional quality in all domains whereas treatment teachers in Location A improved in just one domain. Moreover, only in Location B did the students of treatment teachers make gains in a targeted content area that was significantly larger than that of the students of control teachers.

Our aim is not to defend a position that the two groups of treatment teachers were entirely unique in their learning outcomes but rather to highlight the possibility that differential impact could occur and to consider why this might be the case. One reason for variability across teacher groups might be that they did not experience PD in the same way. We have some evidence that teachers in the two locations did indeed hold very different impressions of the LTG PD workshops.

In their reflections, Location A teachers were markedly more critical than Location B teachers (see Table 8.8). Less than half of the teachers (41%) in Location A described the foundation module (the first five workshops) as a positive experience overall. By contrast, approximately two-thirds of the teachers (67%) in Location B reported having a positive experience. Conversely, 28% of teachers in Location A indicated that the LTG PD was a negative experience, compared to only 2% of teachers in Location B. Another interesting takeaway from these reflections is that teachers can be relatively negative in their perceptions of a PD program but still exhibit gains in the targeted areas.

Table 8.8. Teacher Perception of LTG PD Foundation Module Experience—
Percentage of Teachers (*N* = 41 Treatment Teachers)

General Tone	Location A Treatment (n = 24)	Location B Treatment (n = 17)
Positive	41%	67%
Some positive and some negative	20%	15%
Negative	28%	2%
Can't tell	10%	16%

There are a variety of possible explanations for differences in teachers' perceptions and the nature of the impact of PD across locations. For example, the teachers in Location A had more years of teaching experience in general and more years of experience teaching math. In their reflections, some teachers in Location A expressed disappointment at a perceived lack of challenge related to the mathematical content. Also, because the workshops in Location B were conducted after those in Location A, the facilitator might have made subtle changes that resulted in more productive conversations.

Another possibility is that the community was more positively formed in Location B, leading to stronger engagement in the material. We plan to look carefully at the nature of the facilitation and participation structures within the workshops in both locations to better understand how they differed and the potential connections to learning outcomes.

The results of this efficacy study of LTG PD have broad implications for research on the design and implementation of programs to support teacher learning. Our findings indicate that a highly specified, video-based intervention can be implemented to improve critical aspects of teacher knowledge and (perhaps more importantly) instructional practice, leading to gains in student learning in a targeted content area. It is important to highlight the relatively new instrument used in this study to measure instruction, the math in common protocol.

This instrument targets three broad categories of instructional practices, and in this study it effectively captured shifts in these categories over time, an essential factor for intervention studies such as this one. Further exploration of the math in common protocol appears warranted, especially in longitudinal research on mathematics teacher change.

In addition, the results from this study imply that although teachers in different locations might experience a particular PD program in unique and even unfavorable ways, the program can still have a powerful impact in the desired domains. Context-based effects should be expected at least to some degree given that learning occurs within communities comprised of individuals with unique backgrounds and prior experiences. Future research on professional learning should more deeply investigate implementation differences and teachers' experiences within professional development groups to identify specific factors that may affect their learning.

NOTE

1. The Learning and Teaching Geometry Efficacy Study was supported by the National Science Foundation (NSF award #1503399).

REFERENCES

Academy for Educational Development. (2002). National Writing Project: Final evaluation report. Author: New York, NY.

Akkerman, S. F., & Bakker, A. (2011). Learning at the boundary: An introduction. *International Journal of Educational Research, 50*(1), 1–5.

Avalos, B. (2011). Teacher professional development in teaching and teacher education over ten years. *Teaching and Teacher Education, 27*(1), 10–20.

Baumert, J., Kunter, M., Blum, W., Brunner, M., Voss, T., Jordan, A., . . . & Tsai, Y. M. (2010). Teachers' mathematical knowledge, cognitive activation in the classroom, and student progress. *American Educational Research Journal, 47*(1), 133–180.

Bell, C. A., Wilson, S. M., Higgins, T., & McCoach, D. B. (2010). Measuring the effects of professional development on teacher knowledge: The case of developing mathematical ideas. *Journal for Research in Mathematics Education,* 479–512.

Blomberg, G., Renkl, A., Sherin, M. G., Borko, H., & Seidel, T. (2013). Five research-based heuristics for using video in pre-service teacher education. *Journal for Educational Research Online, 5*(1), 90–114.

Blomberg, G., Sherin, M. G., Renkl, A., Glogger, I., & Seidel, T. (2014). Understanding video as a tool for teacher education: Investigating instructional strategies to promote reflection. *Instructional Science, 42*(3), 443–463.

Bransford, J. D., Brown, A. L., & Cocking, R. R. (2000). *How people learn: Vol. 11.* Washington, DC: National Academies Press.

Brown, J. S., Collins, A., & Duguid, P. (1989). Situated cognition and the culture of learning. *Educational Researcher, 18*(1), 32–42.

Borko, H. (2004). Professional development and teacher learning: Mapping the terrain. *Educational Researcher, 33*(8), 3–15.

Borko, H., Jacobs, J., Eiteljorg, E., & Pittman, M. E. (2008). Video as a tool for fostering productive discourse in mathematics professional development. *Teaching and Teacher Education, 24*, 417–436.

Borko, H., Jacobs, J., & Koellner, K. (2010). Contemporary approaches to teacher professional development. In P. Peterson, E. Baker, & B. McGaw (Eds.), *International Encyclopedia of Education: Vol. 7* (pp. 548–556). Oxford: Elsevier.

Borko, H., Jacobs, J., Seago, N., & Mangram, C. (2014). Facilitating video-based professional development: Planning and orchestrating productive discussions. In Y. Li, E. A. Silver, & S. Li (Eds.), *Transforming mathematics instruction: Multiple approaches and practices* (pp. 259–281). Dordrecht, The Netherlands: Springer International.

Brown, J. S., Collins, A., & Duguid, P. (1989). Situated cognition and the culture of learning. *Educational Researcher, 18*(1), 32–42.

Cobb, P., McClain, K., de Silva Lamberg, T., & Dean, C. (2003). Situating teachers' instructional practices in the institutional setting of the school and district. *Educational Researcher, 32*(6), 13–24.

Cohen, D. K., & Ball, D. L. (1999). Instruction, capacity, and improvement. Consortium for Policy Research in Education, University of Pennsylvania. Retrieved from https://files.eric.ed.gov/fulltext/ED431749.pdf

Desimone, L. M. (2009). Improving impact studies of teachers' professional development: Toward better conceptualizations and measures. *Educational Researcher, 38*(3), 181–199.

Desimone, L. M., Porter, A. C., Garet, M. S., Yoon, K. S., & Birman, B. F. (2002). Effects of professional development on teachers' instruction: Results from a three-year longitudinal study. *Educational Evaluation and Policy Analysis, 24*(2), 81–112.

Firestone, W. A., Mangin, M. M., Martinez, M. C., & Polovsky, T. (2005). Leading coherent professional development: A comparison of three districts. *Educational Administration Quarterly, 41*(3), 413–448.

Franke, M. L., Carpenter, T. P., Levi, L., & Fennema, E. (2001). Capturing teachers' generative change: A follow-up study of professional development in mathematics. *American Educational Research Journal, 38*(3), 653–689.

Gallagher, H. A., Woodworth, K. R., & Arshan, N. L. (2015). *Impact of the National Writing Project's College-Ready Writers Program on teachers and students.* Menlo Park, CA: SRI International.

Gaudin, C., & Chaliès, S. (2015). Video viewing in teacher education and professional development: A literature review. *Educational Research Review, 16*, 41–67.

Gewertz, C. (2013). Teachers say they are unprepared for common core. *Education Week, 32*(22), 1–12.

Greeno, J. G., Collins, A. M., & Resnick, L. B. (1996). Cognition and learning. *Handbook of Educational Psychology, 77*, 15–46.

Grissmer, D. W., Flanagan, A., Kawata, J. H., Williamson, S., & LaTourrette, T. (2000). *Improving student achievement: What state NAEP test scores tell us.* Santa Monica, CA: Rand Corporation.

Grossman, P., Wineburg, S., & Woolworth, S. (2001). Toward a theory of teacher community. *Teachers College Record, 103*(6), 942–1012.

Hatch, T., Shuttleworth, J., Jaffee, A.T., & Marri, A. (2016). Videos, pairs, and peers: What connects theory and practice in teacher education? *Teaching and Teacher Education, 59*, 274–284.

Hill, H. C. (2014). *Mathematical quality of instruction (MQI): 4-point version.* Ann Arbor, MI: University of Michigan Learning Mathematics for Teaching Project.

Ingvarson, L., Meiers, M., & Beavis, A. (2005). Factors affecting the impact of professional development programs on teachers' knowledge, practice, student outcomes, & efficacy. *Education Policy Analysis Archives, 13*(10).

Jacob, R., Hill, H., & Corey, D. (2017). The impact of a professional development program on teachers' mathematical knowledge for teaching, instruction, and student achievement. *Journal of Research on Educational Effectiveness, 10*(2), 379–407.

Jacobs, J., Seago, N., & Koellner, K. (2017). Preparing facilitators to use and adapt professional development materials productively. *International Journal of STEM Education, 4*(1), 30.

Kazemi, E., & Hubbard, A. (2008). New directions for the design and study of professional development: Attending to the coevolution of teachers' participation across contexts. *Journal of Teacher Education, 59*(5), 428–441.

Kennedy, M. M. (2016). How does professional development improve teaching? *Review of Educational Research, 86*(4), 945–980.

Kersting, N. (2008). Using video clips of mathematics classroom instruction as item prompts to measure teachers' knowledge of teaching mathematics. *Educational and Psychological Measurement, 68*(5), 845–861.

Kersting, N. B., Givvin, K. B., Sotelo, F. L., & Stigler, J. W. (2010). Teachers' analyses of classroom video predict student learning of mathematics: Further explorations of a novel measure of teacher knowledge. *Journal of Teacher Education, 61*(1–2), 172–181.

Kersting, N. B., Givvin, K. B., Thompson, B. J., Santagata, R., & Stigler, J. W. (2012). Measuring usable knowledge: Teachers' analyses of mathematics classroom videos predict teaching quality and student learning. *American Educational Research Journal, 49*(3), 568–589.

Koellner, K., & Jacobs, J. (2015). Distinguishing models of professional development: The case of an adaptive model's impact on teacher's knowledge, instruction, and student achievement. *Journal of Teacher Education, 66*(1), 51–67.

Koh, K. (2015). The use of video technology in pre-service teacher education and in-service teacher professional development. In *Cases of mathematics professional development in East Asian countries* (pp. 229–247). Singapore: Springer.

Kutaka, T. S., Smith, W. M., Albano, A. D., Edwards, C. P., Ren, L., Beattie, H. L., . . . & Stroup, W. W. (2017). Connecting teacher professional development and student mathematics achievement: A 4-year study of an elementary mathematics specialist program. *Journal of Teacher Education, 68*(2), 140–154.

Lave, J., & Wenger, E. (1991). *Situated learning: Legitimate peripheral participation.* Cambridge: Cambridge University Press.

Little, J. W. (2003). Inside teacher community: Representations of classroom practice. *Teachers College Record, 105*(6), 913–945.

Loucks-Horsley, S., Stiles, K. E., Mundry, S., Love, N., & Hewson, P. W. (2009). *Designing professional development for teachers of science and mathematics.* Thousand Oaks, CA: Corwin Press.

Major, L., & Watson, S. (2018). Using video to support in-service teacher professional development: The state of the field, limitations and possibilities. *Technology, Pedagogy and Education, 27*(1), 49–68.

McCallum, W. (2011). Gearing up for the common core state standards in mathematics. *Draft report of the policy workshop.* http://internet.math.arizona.edu/~ime/2010–11/2011_04_01_IME_PD.pdf

Miller, K., & Zhou, X. (2007). Learning from classroom video: What makes it compelling and what makes it hard. In R. Goldman-Segal & R. Pea (Eds.), *Video research in the learning sciences* (pp. 321–334). Hillsdale, NJ: Lawrence Erlbaum and Associates.

Penuel, W. R., Fishman, B. J., Yamaguchi, R., & Gallagher, L. P. (2007). What makes professional development effective? Strategies that foster curriculum implementation. *American Educational Research Journal, 44*(4), 921–958.

Perry, R. R., Finkelstein, N. D., Seago, N., Heredia, A., Sobolew-Shubin, S., & Carroll, C. (2015). *Taking stock of common core math implementation: Supporting teachers to shift instruction.* San Francisco, CA: WestEd.

Perry, R., Seago, N., Burr, E., Broek, M., & Finkelstein, N. (2015). *Classroom observations: Documenting shifts in instruction for districtwide improvement.* San Francisco, CA: WestEd.

Putnam, R. T., & Borko, H. (2000). What do new views of knowledge and thinking have to say about research on teacher learning? *Educational Researcher, 29*(1), 4–15.

Raudenbush, S. W., Bryk, A. S., Cheong, Y. F., Congdon, R., & Du Toit, M. (2011). *Hierarchical linear and nonlinear modeling (HLM7).* Lincolnwood, IL: Scientific Software International.

Roth McDuffie, A., Drake, C., Choppin, J., Davis, J. D., Magaña, M. V., & Carson, C. (2017). Middle school mathematics teachers' perceptions of the common core state standards for mathematics and related assessment and teacher evaluation systems. *Educational Policy, 31*(2), 139–179.

Saderholm, J., Ronau, R., Brown, E. T., & Collins, G. (2010). Validation of the diagnostic teacher assessment of mathematics and science (DTAMS) instrument. *School Science and Mathematics, 110*(4), 180–192.

Santagata, R., Kersting, N., Givvin, K. B., & Stigler, J. W. (2010). Problem implementation as a lever for change: An experimental study of the effects of a professional development program on students' mathematics learning. *Journal of Research on Educational Effectiveness, 4*(1), 1–24.

Schoenfeld, A. H. (2013). Classroom observations in theory and practice. *ZDM, 45*(4), 607–621.

Schoenfeld, A. H., Floden, R. E., & the Algebra Teaching Study and Mathematics Assessment Project. (2014). *The TRU math scoring rubric.* Graduate School of

Education, University of California, Berkeley, and College of Education, Michigan State University. Retrieved from http://ats.berkeley.edu/tools.html.

Scribner, J. P. (1999). Professional development: Untangling the influence of work context on teacher learning. *Educational Administration Quarterly, 35*(2), 238–266.

Seago, N., Jacobs, J., Driscoll, M., Callahan, P., Matassa, M., & Nikula, J. (2017). *Learning and teaching geometry: Video cases for mathematics professional development, grades 5–10.* San Francisco, CA: WestEd.

Seago, N., Jacobs, J., Driscoll, M., Nikula, J., Matassa, M., & Callahan, P. (2013). Developing teachers' knowledge of a transformations-based approach to geometric similarity. *Mathematics Teacher Educator, 2*(1), 74–85.

Seago, N. M., Jacobs, J. K., Heck, D. J., Nelson, C. L., & Malzahn, K. A. (2014). Impacting teachers' understanding of geometric similarity: Results from field testing of the learning and teaching geometry professional development materials. *Professional Development in Education, 40*(4), 627–653.

Sherin, M., Jacobs, V., & Philipp, R. (Eds.). (2011). *Mathematics teacher noticing: Seeing through teachers' eyes.* New York, NY: Routledge.

Sherin, M., & van Es, E. A. (2009). Effects of video club participation on teachers' professional vision. *Journal of Teacher Education, 60*(1), 20–37.

Star, S. L., & Griesemer, J. R. (1989). Institutional ecology, 'translations' and boundary objects: Amateurs and professionals in Berkeley's Museum of Vertebrate Zoology, 1907–39. *Social Studies of Science, 19*(3), 387–420.

Sztajn, P., Marrongelle, K. A., Smith, P., & Melton, B. L. (2012). *Supporting implementation of the common core state standards for mathematics: Recommendations for professional development.* Raleigh, NC: William and Ida Friday Institute for Educational Innovation, North Carolina State University College of Education.

Taylor, J. A., Roth, K., Wilson, C. D., Stuhlsatz, M. A., & Tipton, E. (2017). The effect of an analysis-of-practice, videocase-based, teacher professional development program on elementary students' science achievement. *Journal of Research on Educational Effectiveness, 10*(2), 241–271.

Teuscher, D., Tran, D., & Reys, B. J. (2015). Common core state standards in the middle grades: What's new in the geometry domain and how can teachers support student learning? *School Science and Mathematics, 115*(1), 4–13.

Tran, D., Reys, B. J., Teuscher, D., Dingman, S., & Kasmer, L. (2016). Analysis of curriculum standards: An important research area. *Journal for Research in Mathematics Education, 47*(2), 118–133.

van Es, E. A. (2012). Using video to collaborate around problems of practice. *Teacher Education Quarterly, 39*(2), 103–116.

van Es, E. A., & Sherin, M. G. (2008). Mathematics teachers' "learning to notice" in the context of a video club. *Teaching and Teacher Education, 24*(2), 244–276.

van Es, E. A., & Sherin, M. G. (2010). The influence of video clubs on teachers' thinking and practice. *Journal of Mathematics Teacher Education, 13*(2), 155–176.

Wayne, A. J., Yoon, K. S., Zhu, P., Cronen, S., & Garet, M. S. (2008). Experimenting with teacher professional development: Motives and methods. *Educational Researcher, 37*(8), 469–479.

Webster-Wright, A. (2009). Reframing professional development through under-standing authentic professional learning. *Review of Educational Research, 79*(2), 702–739.

Wei, R. C., Darling-Hammond, L., & Adamson, F. (2010). *Professional development in the United States: Trends and challenges, vol. 28.* Dallas, TX: National Staff Development Council.

Yoon, K. S., Duncan, T., Lee, S. W.-Y., Scarloss, B., & Shapley, K. (2007). *Reviewing the evidence on how teacher PD affects student achievement.* Washington, DC: U.S. Department of Education.

Chapter 9

A Complex Adaptive Model of Algebra Professional Development

Sarah Smitherman Pratt and Colleen McLean Eddy

INTRODUCTION

The researchers of this chapter present their study that used the structure of complex adaptive systems to analyze an algebra professional development (PD) model to determine to what degree, if any, PD met the emerging needs of algebra teachers as they engaged in inquiry-based instruction. The method in this study used the three processes within complexity theory: critical realism (CT-CR) of system mapping; extended case studies; and process tracing to analyze iterations of PD over nine years.

SYSTEMS PERSPECTIVE FOR PD

The perspective of complex dynamical systems was at the forefront of our adaptive PD model for algebra teachers. Each facet of the model addressed and continues to address the complexities of learning and teaching in the context of teaching mathematics to adolescent learners with diverse needs. We outline how complex dynamical systems, as situated within complexity theory (CT), and education research informs the structure and content of PD.

In this study we drew on "dynamic connectivities between and among elements and structures that continuously change over time" (Cochran-Smith, Ell, Ludlow, Grudnoff, & Aitken, 2014b, pp. 28–29) to analyze how an algebra PD model for teachers continued to be effective in transforming knowledge and practices of the participants as it evolved. We offer this as an opportunity to seek "complex, contingent, and multiple causes while at the same time avoiding being reductionist, linear, or piecemeal" (p. 21). Algebra is used here to describe the content covered in a first-year course as outlined by Eddy et al. (2015).

As PD facilitators, we developed and implemented a year-long PD program with iterations over nine years, and the findings provided describe aspects of the structure and content by utilizing the CT-CR research design. Analyses of these findings demonstrated the importance of maintaining complex and contingent notions of agency and responsibility that depend on deep understanding of the local (e.g., initial conditions, sequences, and transformative events) linked to larger understandings of processes and outcomes at various systems levels that are widely variable but potentially explicable.

PURPOSE OF THE STUDY

Current research on effective PD models shows that PD models have a longer and more lasting impact on teacher professional growth when they are highly adaptive (Koellner & Jacobs, 2015; Lewis & Perry, 2015). The purpose of this study was to use the structure of complex adaptive systems to analyze an algebra PD model to determine to what degree, if any, PD met the emerging needs of algebra teachers as they engaged in inquiry-based instruction.

The research question for this study was, In what ways does an algebra PD model adapt as the needs and educational contexts of algebra teachers shift? To answer this question, the researchers analyzed the data based on CT-CR research design (Cochran-Smith et al., 2014b) in order to examine the changes that occurred. This method highlighted what aspects of the model remained and what aspects were added, altered, or removed over the duration of the nine years.

Before reporting our findings, we define key aspects of effective PD. We then describe what is meant by *complex* in the context of CT and education research. Following this we provide the methods used to collect and analyze our findings. We conclude with a discussion of the themes that emerged from our analysis and implications for future research in teacher PD.

DEFINING EFFECTIVE PD

PD is intended to improve the practice and effectiveness of teaching. Hirsh (2009) defined PD as a "comprehensive, sustained, and intensive approach to improving teachers' and principals' effectiveness in raising student achievement" (p. 12). In an extensive meta-analysis of PD, Hattie (2008) reported on over 800 research studies related to student achievement. He concluded that teachers should obtain an effect size greater than $d = 0.40$ to be considered better than average based on the findings from the different studies. Further, he found that out of the six categories for influencing student achievement, teaching had the most meta-analysis studies showing an effect size greater than $d = 0.40$.

Discussion of why this occurred led Hattie to conclude that models of effective PD must, at a minimum, occur across an entire year and be focused on a specific curriculum that includes active learning, teachers participating collectively, and an alignment with the goals of the school and their students (Blank, de las Alas, & Smith, 2007; Desimone, 2009; Hirsh, 2009).

In mathematics teacher education, Koellner, Jacobs, and Borko (2011) demonstrated that what is "essential in preparing leaders to implement high-quality mathematics PD" are "(1) fostering a professional learning community, (2) developing teachers' mathematical knowledge for teaching, and (3) adapting PD to support local goals and interests" (p. 116). In this study we drew on Koellner, Jacobs, and Borko's (2011) assertion, which resonates with Hattie's (2008) meta-analysis of effective PD, to show that complex, adaptive PD across its iterations continues to prepare algebra teachers provided it is what Koellner, Jacobs, and Borko (2011) would consider high-quality PD.

CT and Education Research

Part of our model was the complex adaptive nature of the model across iterations. The term *complex* is drawn from CT, which is based on non-linear, dynamical structures of living systems (Doll, 1993; Kaufmann, 1996; Mason, 2008; Morin, 2008). Systems that adapt and create emergent properties maintain vitality while systems that remain fixed and inert eventually become extinct (Bateson, 1979/2002; Doll, 1993; Maturana & Varela, 1987). We embraced complexity from this perspective as we posited what it meant for teachers to engage in algebra PD.

William Doll (1993, 2005, 2012) and colleagues (Doll & Gough, 2002; Doll, Fleener, St. Julien, & Trueit, 2005) focused on interrelationships and connectivities. Doll (1993) highlighted the importance of inquiry, indeterminacy, and reflection in the acts of learning and teaching (Pratt, 2018). These tenets, when viewed from a complex dynamical sense, supported the hypothesis that interactions among teachers and students as a community would lead to different pathways and explorations as meanings can be co-created.

Davis and Simmt (2003) outlined important aspects of communities that display a complex dynamic. They focused on five conditions not to separate them distinctively but rather to discuss them in relation to each other and to the system as a whole: "(a) internal diversity, (b) redundancy, (c) decentralized control, (d) organized randomness, and (e) neighbor interactions" (p. 147). These were described as *interdependent conditions* in reference to the "global properties of a system and to the local activities of agents within a complex system" (p. 147). The logic of both/and has been maintained as a distinction in complexity thinking; considerations for how each are true come into play in the decisions and interactions of the system (for more on this, see Pratt [2008a, 2008b]).

Later, Davis and Simmt (2016) built on these conditions and elaborated on complex perspectives for mathematics learning as related to education research in mathematics education. They defined the traits of a complex learning form/phenomenon/entity. These include

- co-dependently arising with the world in the co-implicated interactions of multiple agents/systems
- being typically conceived/perceived and characterized as a body and/or in embodied terms
- manifesting features and capacities that are not observed in constituting agents/systems,
- maintaining itself over some period of time
- evolving in response to both internal and external dynamics in manners that are better described in terms of adequacy/sufficiency (i.e., fitness among subagents and between agent and environment) rather than optimality/efficiency (i.e., match between internal and external; p. 423).

We adopted Davis and Simmt's (2016) definition of a complex form of learning to describe our algebra PD model, and the model was used as a case study with all of the interactive agents within the system as co-constructing and co-evolving. This study contributes to education research by describing an extended case study to "simultaneously highlight common ground and important divergences" (p. 424). In the process of tracing the elements within the system, the model depicts an evolving system.

In defining the PD model as a system, we maintained the perspective of a "proposed adaptability continuum" in which we "consider the synergistic whole of the PD model, recognizing the unique juxtaposition of the goals, expectations, and contextual elements that together form the core of the model" (Koellner & Jacobs, 2015, p. 52). Complex conversations in education can produce emerging new ideas that transform all participants (Pratt, 2008a; Trueit, 2012).

METHODS OF CT-CR

This study drew on CT as described above both as a framework and as a methodology. Using CT as a framework, we worked with teachers utilizing complex conversations (Pratt, 2008a) to inform, transform, and re-form our collective professional learning community. Using CT as a methodology, we recursively analyzed our instructional model and strategies to adapt and change as we considered implementations in future iterations. This is what we mean by complex adaptive PD.

Cochran-Smith et al. (2014a, 2014b, 2016) suggested that empirical research studies embrace a CT-CR research design and consider three analytic processes: (1) system mapping, (2) extended case studies, and (3) process tracing.

Cochran-Smith et al. (2014a) argued that when CT is combined with CR, the method employed provides a unique perspective "because it offers a theoretical framework that preserves wholes, privileges interactions and interdependencies, and expects surprising outcomes" (p. 33), and when CT and CR are paired, the data can show how complex systems shift and change over time. Cochran-Smith et al. emphasized that "a central aspect of complex systems is the interactions, interrelationships, and interdependencies of elements rather than discrete elements or disconnected parts of a process" (p. 28).

In line with a complex perspective, these analytic processes are not mutually exclusive nor do they suggest selecting only one. Instead, they "can be used in combination with each other and/or with other existing qualitative, quantitative, and mixed methods approaches to lead to promising new lines of research in teacher education" (p. 28). We take the suggestions of Cochran-Smith et al. (2014a) to analyze the data we collected using the CT-CR research design and the three analytic processes.

System Mapping

System mapping, "which lays out the general landscape of a complex system, including its major elements and structures, its interdependencies and overlapping areas, and its ambiguous borders" (Cochran-Smith et al., 2014b, p. 29), is used to describe the context and focus of our PD as it continues to shift and morph over time. This relies on the perspective that organizations should be regarded as complex and flexible systems that operate not in terms of simple relationships between the discrete pieces of a system but in terms of dynamic connectivities between and among elements and structures that continuously change over time (pp. 28–29).

In this study, PD is the system, and the components of the system include the PD instructors, the participants, the algebra content, and the teaching contexts of the participants' schools. The method used in this study relies on "identifying and understanding the multiple, contingent, and complex causes of particular outcomes within and across cases, with a focus on mechanism- and process-based explanations" (Cochran-Smith et al., 2014b, p. 31). Table 9.1 provides an overview of the elements of the PD and how it shifted and morphed over nine years.

Table 9.1. **System Mapping of XTreem Math over Nine Years**

	Inquiry-Based Instruction (Algebraic Habits of Mind & 5E Lessons)	*Short-Cycle Formative Assessment*	*Orchestrating Productive Discussions*	*Lesson Study with Open Approach (LSOA)*
Year 1	Modeled during PD			
Year 2	Modeled during PD			
Year 3	Modeled during PD	Learning progressions		
Year 4	Modeled during PD	Learning progressions		
Year 5	Modeled during PD	*AssessToday* observation protocol		
Year 6	Modeled during PD	*AssessToday* observation protocol		
Year 7	Modeled during PD	*AssessToday* observation protocol		
Year 8	Modeled during PD	*AssessToday* observation protocol	Introduced	Pilot conducted over one round
Year 9	Modeled during PD	*AssessToday* observation protocol during LSOA lesson	Monitoring Form Incorporated in LSOA lessons	Two rounds conducted

Extended Case Studies

By collecting PD model data over time, our extended case studies aim "to shift the focus away from the knowledge and skill of individual teacher candidates and toward the ways that individuals' experiences and performances are shaped by complex practice environments and organizations" (Cochran-Smith et al., 2014b, p. 31). The information gleaned by each iteration of the PD shaped and continues to shape future decisions regarding the framework and implementation of the PD with new cohorts.

Designs of extended case studies allow for a description of how "relationships are critical and nonlinear, dynamics are unpredictable, and interdependencies exist across the boundaries and levels of the system" (Cochran-Smith et al., 2014b, p. 30). Anderson, Crabtree, Steele, and McDaniel (2005) suggest that the method of extended case studies, "because of the coevolutionary nature of the system," calls us to

> pay more attention to the interdependencies across the boundaries of systems. Traditionally, case studies bound the case and then study phenomenon within the boundary. Complexity science suggests that important insights can be gleaned by studying the behavior that occurs at and across the boundaries that define the case. (p. 674)

Taking this approach, "with extended case studies, there is more emphasis on interrelationships, flows, and exchanges" (Cochran-Smith et al., 2014b, p. 31). This study was an extended case study of an algebra PD program. The participants were teachers from an urban setting who taught algebra in middle or high school. Figure 9.1 shows the sequence of years with PD as the case study. The figure also includes process tracing across iterations.

●	○	●	○	●	●	○	●	●
Year 1 (2008-09)	Year 2 (2009-10)	Year 3 (2010-11)	Year 4 (2011-12)	Year 5 (2012-13)	Year 6 (2013-14)	Year 7 (2014-15)	Year 8 (2015-16)	Year 9 (2016-17)
Project #1	*Project #2*			*Project #3*		*Project #4*		*Project #5*

● - Preparation and implementation of key aspects in the PD

○ - Recognition for an adaption of key aspects in the PD

Figure 9.1. Iterations of the PD across nine years. *Authors*: Sarah Smitherman Pratt and Colleen McLean Eddy

Process Tracing

In CT-CR, the method is not to be interpreted as a cause-effect set of discrete relations but instead should be considered in the moment-by-moment changes as actions and reactions to different aspects and participants in the system.

With this perspective, process tracing is included in CT-CR to "track the key causal processes and mechanisms that support, constrain, amplify, or diminish" (Cochran-Smith et al., 2014b, p. 32) components within the system.

By using process tracing throughout the adaptations, we emphasize "identifying and understanding the multiple, contingent, and complex causes of particular outcomes within and across cases, with a focus on mechanism- and process-based explanations" (Cochran-Smith et al., 2014b, p. 31). When we use the PD model as the case study, by way of the processes (see Figure 9.1) that are analyzed we seek to identify and explore the "multiple, contingent, and complex causes of particular outcomes within and across cases, with a focus on mechanism- and process-based explanations" (p. 31).

We integrated these three analytic processes to describe how our PD model continues to adapt as a complex framework that empowers teachers to be their own agents of change. Findings from the research study that emerged from the integration of these processes are described by key aspects of PD in the following section.

Data Collection

The data collection began in 2008 and has continued to date. Institutional Review Board approval was obtained for each iteration of the PD. Data collected included teaching observations, post-observation conferences, written reflective prompts, high-stakes testing results, thinking sheets completed by individual participants, monitoring forms completed by groups, and statements made during PD sessions.

These data helped to inform key aspects of PD analyzed through the lens of CT-CR. By drawing on CT-CR, we shifted from analyzing the individual changes of participants and single-year results toward patterns that connected iterations of the studies. This approach maintained a CS perspective of the contributors and factors involved in PD.

Participants

Participants in the study were recruited to apply for grant-funded algebra PD. Selection of algebra teachers was prioritized according to those who had one or more of the following: (1) five years of teaching experience or less, (2) alternative certification, (3) a certification not in mathematics, and (4) a teaching position in a school where 50% or more of the students were of low socioeconomic status. Participants were mathematics teachers who were teaching middle grade mathematics or algebra at the time of application.

Each iteration included about 20 participants. Each held at least a bachelor's degree and was a certified teacher. Their ages ranged from 25 to over 60 years old and their experience from one to over 20 years of teaching.

FINDINGS

The integration of data from years of implementation generated descriptions that, using a systems perspective for this interpretation, showed initial conditions and how potential small changes could make a large difference. The key aspects of PD found in the analysis centered on inquiry-based instruction, short-cycle formative assessment, orchestrating productive discussions, and lesson study with open approach. Below is a description of each aspect of the PD.

Inquiry-Based Instruction

Initially, the two components of the PD sessions centered on building teachers' knowledge of teaching by engaging in active learning that highlighted inquiry-based instruction (in the form of the 5E lesson model as defined by Bybee et al. [2006]) and in building teachers' mathematical knowledge using Driscoll's (1999) *Fostering Algebraic Habits of Mind*. Additionally, participants traveled to places such as an amusement park to engage in hands-on, authentic applications of algebraic concepts.

The inquiry-based lessons highlighted a constructivist epistemology (Bybee, 2014; Bybee et al., 2006) by creating a learning trajectory for the participants to follow. Models of inquiry-based instructional plans were provided and implemented during the sessions to demonstrate effective strategies for teaching from an inquiry-based approach as well as to allow the participants to obtain first-hand experience regarding the lessons. Participants were also given materials to use for implementing these lessons in their own classrooms as well as online access to all the instructional plans. In subsequent sessions, participants discussed how their own teaching of the lessons transpired and reflected on their own learning from the experience.

The work of Driscoll (1999) was also integrated into the instructional sessions. Specifically, tasks that focused on ways in which to engage students through effective questioning and that were rich with the potential to allow students to demonstrate their understanding of algebraic concepts were incorporated to complement the inquiry-based lessons. The tasks were designed to attend to algebraic concepts through open-ended problem solving.

Drawing from their participation in the inquiry-based lessons, from tasks framed around algebraic habits of mind, and from experiential learning that included trips to other locations of learning, some of the participants created their own inquiry-based lessons, and from these some were selected to be used in subsequent years in the program. These lessons were examples of how PD impacted teaching practices and pedagogical strategies.

Short-Cycle Formative Assessment

The PD facilitators noted that inquiry-based lessons allowed teachers to capture the understanding of their students, which led to formative assessment strategies being incorporated into the program. Subsequently, as an adaptation in Year 3 of the program, PD facilitators added to the sessions a focus on the learning of students in the form of learning progressions (Popham, 2008). This evolved into providing participants with information regarding aspects of formative assessment and strategies to use such as checklists, which they in turn implemented in their classroom instruction to inform them of students' understanding of a concept. Additionally, participants analyzed sample student work during PD sessions to discuss interpretations of student understanding of the concept and possible instructional adjustments.

As the project progressed through Year 3 and Year 4, we recognized a need for the participants to be able to more effectively implement formative assessment strategies. As a result, we developed an instrument that would be useful for the participants, namely the AssessToday observation protocol (Eddy & Harrell, 2013; Eddy, Harrell, & Heitz, 2017), which includes seven dimensions for implementing short-cycle formative assessment strategies and four levels of proficiency in the implementation of a single lesson.

The seven dimensions of AssessToday include learning target, question quality, nature of questions, self-evaluation, observation of student affect, instructional adjustment, and evidence of learning. (A more detailed description of each of the dimensions can be found in Heitz [2013].) These constructs of short-cycle formative assessment are not meant to be mutually exclusive for they interconnect in meaningful ways for the teachers who use them. Since Year 5, all participants in the project have been observed using the Assess-Today observation protocol. This protocol is described in more detail in Eddy et al. (2017).

Following the observations, post-observation conferences are conducted in small groups for the benefit of all the participants. Each conference begins with a prompt for each participant to share one dimension they believe was effective and one to focus on for future improvement. The feedback and conversations around the seven dimensions allow all who are engaged in the debriefing session to reflect on their own uses of short-cycle formative assessment strategies and how they can improve upon their own implementations.

The project director who conducts the observations guides the conversation to ensure both that all participants share and the focus remains on short-cycle formative assessment strategies.

One participant (we have used pseudonyms) in a recent session reflected on the experience by stating that she is now

> more aware of student thinking and the way students respond. Changing the way I question my students has had an impact on the way my students also respond. I can have a better view of how well students are grasping the mathematic concepts. I also have built solid relationships with my students so they feel comfortable answering questions whether correct or not. (Andrea, Year 8)

Andrea is one example but she reflects the sentiments of many participants who participated in the observations and post-observation conferences. Participants continue to share that this experience is much more useful than evaluations performed by school administrators because they provide detailed, specific feedback and a focus for areas in which to improve, and this generates opportunities to learn and grow professionally. AssessToday continues to be used to formatively assess each teacher's use of short-cycle formative assessment strategies.

By using the seven dimensions of AssessToday as the framework for discussing short-cycle formative assessment, participants can reflect on how intentionally listening and responsively questioning become integrated with instructional time. One participant described the impact in this way:

> The greatest impact is that I have implemented the [required] time needed for allowing time for my students to actually talk so they can get behind their thinking and have time to formulate their concepts. In my allowing this it will afford me the opportunity of also seeing the areas in need of reteach immediately versus awaiting until after the data from a test. (Jesse, Year 8)

This reflection aligns with the dimension of the nature of questioning, which is an integral component to using just-in-time instructional adjustments.

More holistically, as the instructors of the PD reflected on the patterns that emerged from the AssessToday observations, conversations arose around the need for participants to improve the implementation of inquiry-based instruction based on the observation data. This led to another adjustment of the PD.

Orchestrating Productive Discussions

A major consideration for adapting key aspects of the PD was the recognition that participants struggled to understand their role as a facilitator rather than a lecturer during an inquiry-based lesson. To address this, the five practices developed by Stein, Engle, Smith, and Hughes (2008) were offered as a framework the participants could use to imagine how to engage differently. This framework was provided during Year 7 but it was not yet a key aspect of the PD.

In Year 8 participants were given *Five Practices for Orchestrating Productive Discussions in Mathematics* (Smith & Stein, 2011), which was written specifically for teachers. In this book, the authors provide a series of steps to facilitate meaningful discussions in mathematics classrooms to move the approach away from students performing a show-and-tell toward the teacher intentionally determining who should present and in what order, with a trajectory in mind that will allow all students to benefit from the conversations and optimize their learning.

By using the five practices of anticipating, monitoring, selecting, sequencing, and connecting (Smith & Stein, 2011), the teacher can intentionally listen to students during a mathematical task and responsively question as she monitors. Furthermore, as she selects and sequences how the students will share their work, the teacher continues to intentionally listen by responsively questioning the whole class regarding what is being shared.

As PD instructors, we agreed with the five practices and wanted participants to be aware of them as well as model them in their sessions. We recognized that we should take into consideration how to introduce them into the PD by keeping in mind what Hattie (2008) articulated as essential for effective PD, that teachers should participate collectively and actively. Furthermore, as Koellner, Jacobs, and Borko (2011) asserted, we wanted to promote a professional learning community and at the same time adapt the PD to support the goals and interests of our partnering district.

We saw the need for an intermediary step between implementation of short-cycle formative assessment strategies during inquiry-based instruction and orchestration of productive discussions in the classroom. We researched how other PD models could enable teachers to make the transition from participating in the PD to using the strategies with students, and we decided lesson study with open approach (LSOA) provided teacher support in a community environment while teaching to students in the classroom.

LSOA

LSOA, as defined by Inprasitha (2010), incorporates traditional elements of lesson study—plan, teach, and refine—with the added specificity of incorporating open-ended tasks during instruction. Participants collaborate through all elements of lesson study, and research has shown that this develops both depth of knowledge of content and pedagogy (Lewis & Perry, 2015, 2017;

Chinnappan & Cheah, 2012). The open-ended task is selected based on its content focus and potential for a rich discussion.

Koellner and Jacobs (2015) describe lesson study as one of the most adaptive PD models on a continuum from highly adaptive to highly specified. Avalos (2011) reviewed ten years of manuscripts on PD and found that five themes emerged, including mediation. Lesson study is a strategy of teacher co-learning that falls within mediation. Lesson study provides a necessary as well as effective tool in PD allowing teachers to observe and give feedback. As proponents of highly adaptive PD, we incorporated LSOA initially as a pilot in Year 8 (Project 4) and then fully implemented it in Year 9 (Project 5) to impact participants' learning in meaningful ways and to facilitate participants integrating the five practices into their classroom.

Specific to our PD model, during the planning stage participants were required to prepare measures for short-cycle formative assessment (Wiliam & Thompson, 2007) by anticipating possible student responses to an open-ended task, determining how the task would be monitored, selecting and sequencing how student work would be discussed, and connecting the student work according to the purpose of the lesson during the discussion (Smith & Stein, 2011).

Open-ended tasks have been embedded in PD since its initial implementation through *Fostering Algebraic Habits of Mind* (Driscoll, 1999) to engage teachers in thinking of different approaches to mathematical tasks, which then allowed them to focus on student thinking and learning when these tasks were used in their classroom. The adaptation is a direct connection between the open-ended task and the construction of a monitoring form to be used during instruction.

During the implementation of the lesson, the teacher who self-selects to implement the research lesson is observed using the AssessToday observation protocol. The monitoring form is used by the teacher to examine student work and facilitate whole-class discussion once the groups have completed the open-ended task. The other teacher participants are assigned a focus for their observation of the lesson ranging from one student's engagement to one group's interactions to the teacher's questions.

The information gleaned during the lesson is shared in the next phase of LSOA, which is the debriefing session. As LSOA was implemented in Year 9, the need for a set of protocols became apparent. An observation protocol and a debriefing protocol were developed and guided the process of conducting the lesson study round. Appendix A lists the two protocols.

The debriefing session is an opportunity to consider if the research lesson was effective by asking all the participants, "What do you really know about what the students know about the concept, and how do you know it?" Beginning with the participant who taught the lesson, the participants reflect on this question. The information provided by the teacher, who is reflecting on pedagogical decisions and actions,

> represents a complementary approach to finding out what is valued and considered important by the teacher. Such decisions are made, and actions demonstrated in various aspects of our professional lives as teachers, from lesson planning to lesson execution and to assessment. (Kadroon & Inprasitha, 2013, p. 101)

When the group debriefs, the focus of examining the effectiveness of the research lesson shifts the attention from the teacher to the students. However, the teacher plays a role and can reflect on considerations for future instruction not just in terms of the research lesson but also on a daily basis. The fact that the teacher is not being told by an evaluator but rather provided an opportunity to reflect metacognitively means that the learning is in the hands of the teacher. Such is the power of lesson study.

During Year 9, the PD facilitators recognized another benefit to incorporating LSOA and AssessToday observations together. When the participants are provided feedback that is offered to each person observed, they can all meaningfully reflect on the feedback since all were present for the teaching. By incorporating a shared observation, the conversation contains more insight and brings more depth to considerations for implementing formative assessment strategies.

DISCUSSION

As teacher educators, our work is to examine in what ways we can encourage in-service teachers to develop a pedagogy that fosters student mathematical thinking and self-efficacy. In recent years, researchers have shown that teacher use of formative assessment strategies is the most effective approach to meeting the needs of all learners (e.g., Black, 2004; Black & Wiliam, 1998; Rodriguez, 2004). The implementation of formative assessment, however, continues to be an elusive and ambiguous ideology. Several researchers have developed models for implementing formative assessment strategies including Duckor (2014) and Fennell, Kobett, and Wray (2017).

Formative Assessment

As mentioned earlier, Eddy and Harrell (2013) developed a short-cycle formative assessment observation protocol, AssessToday, to provide a more targeted approach to PD that allows for flexibility within various school districts and that is based on research regarding effective, short-cycle, formative assessment strategies. AssessToday was included in this study to impact how participants intentionally listen to and responsively question students. The observation protocol was designed as a formative assessment of the use of each teacher's formative assessment strategies.

The study draws on LSOA (Inprasitha, 2010, 2015) to create professional learning communities that provide opportunities to collaboratively implement inquiry-based instruction that will facilitate meaningful and productive discussions. As the teachers participate in rounds of lesson study, they are observed using AssessToday and participate in post-observation conferences as a group so that all benefit from the feedback.

This adapted PD model includes the goal of developing participants' ability to intentionally listen and responsively question. Our PD model involves the building of these skills in such a way that the participants become engaged in considering how teachers can engage in meaningful and rich conversational moments of learning while not doing all the talking. This was a difficult shift in pedagogy for many of the participants who were comfortable in speaking but unsure how to relinquish control of classroom discussions.

Davis (1997) argues,

> The teacher can be directive within a dialogue with learners. This then, is another important element of a listening posture: In defining one's role more in terms of listening than of telling or facilitating, the teacher need not abandon his or her responsibility to the mandated curriculum guidelines. On the contrary, listening engenders a particular response-ability. (p. 373)

While the teacher may not be speaking, the actions of the teacher are just as rigorous for "listening is an active, and not merely a passive, receptive process. Listening takes attention and effort" (Burbules & Rice, 2010, p. 2878).

Questioning and listening are embedded with both/and logic (Pratt, 2008b). One is not mutually exclusive of the other if they are to inform a living system. Nor is one the causal effect of the other. Instead, they dynamically interplay with each other to create meaningful conversations.

Intentional Listening

In coordination with the questioning during instruction, listening to responses in order to understand and adjust instruction is crucial. Empson and Jacobs (2008) suggest that "learning to listen to children's mathematics requires the development of an interrelated and situated set of skills for attending to children's mathematics in complex instructional environments" (p. 267).

The act of listening to students' responses to questions takes practice and a transformation of pedagogical content knowledge. Teachers have to recognize what students may or may not understand and bring all these aspects of the dialogue into the classroom discourse. This means "eliciting, interpreting, and following up on students' reasoning in the moment, in a fashion that values students' ideas as objects of inquiry" (Maskiewicz & Winters, 2012, p. 433). Valuing students' ideas encourages students to share more and to take risks because their thoughts are respected and considered.

Being attentive to students' actions and shared ideas in nonverbal ways is just as important. Placing manipulatives in their hands to elicit their thoughts allows students to express what they are thinking even if they cannot metacognitively express their conceptual understandings. When teachers are

> interested in "making sense of the sense they're making" [then] student talk is thus understood as a means of making public one's emerging ideas, or representing them, and in that re-presentation, taking advantage of an opportunity to formulate and reformulate. The use of the manipulative materials is critical here: they serve as a commonplace for learners to talk about ideas, enabling the processes of re-presentation and revision." (Davis, 1997, p. 365)

Responsive Questioning

Responsive questioning involves the attunement (Aoki, 1993/2005) of the teacher to what the students are thinking and sharing. As teachers engage in formatively assessing their students, they can adapt their instruction if they ask questions that will effectively elicit student thinking. Active listening is just as significant, for the answers to questions will not impact instructional adjustments if the teacher does not listen and consider what students are saying.

Listening and understanding are never simply passive acts of reception; they are "active acts of reconstructing or of interpreting" (Huebner, 1977/1999, p. 264). As Huebner (1963/1999) states, it is the listener who "establishes the climate for conversation" (p. 79). Our goal in this PD model was for participants to establish a climate of conversation in which genuine exchanges occurred when the teacher focused on "making sense of the sense that the students are making" (Davis, 1997, p. 365).

Implications

As researchers and teacher educators, we strive to empower the teachers with whom we work and offer opportunities for professional growth as they engage in activities that empower them. We use CT-CR to describe and analyze how we continue to adapt and shift our model as the needs of our participants change over time. We use an extended case study of PD in order to depict in what ways these changes occur and to what level there is an impact.

This study adds to current research related to teacher education and PD by providing a different way to examine the effectiveness of a PD model. As other PD models are implemented and adapted across iterations, they could also be examined using this approach. CT-CR allows for a description of momentary process-changes and also larger structural changes across iterations. The focus is not on the individuals but rather on the system.

CONCLUSION

We argue that when teacher educators report their findings from their PD, CT-CR provides a useful framework for analyzing the effectiveness of the PD. The systems approach gives voice to the diverse aspects found within the system. Additionally, describing the components does not reduce the system but rather looks at the system from multiple perspectives. Complexity theory: critical realism is not an approach to determine what will work and what can be replicated for other systems; rather, it is an approach that allows for diversity and variance within a system that will change and affect how the system adapts across time.

ACKNOWLEDGMENTS

This material is based on work supported by Teacher Quality Grants under the No Child Left Behind Act. Any opinions, findings, interpretations, conclusions, or recommendations expressed in this material are those of the authors and do not necessarily represent views of the Texas Higher Education Coordinating Board.

APPENDIX A: PROTOCOL FOR PREPARING, IMPLEMENTING, AND DEBRIEFING THE RESEARCH LESSON FOR LESSON STUDY WITH OPEN APPROACH (LSOA)

Preparing for the Research Lesson

Group Decisions

1. Select a lesson plan to be used as the research lesson.
2. Determine what open task will be used during the lesson.
3. Create a monitoring form based on the open task selected.
4. Assign roles for the first iteration of the research lesson.
5. Decide who will provide the following on the day of the lesson study round:
   - manipulatives needed for the lesson
   - handouts for the students
   - seating chart for the classes being observed
   - name tags or name cards for the students in the classes being observed
   - copies of the monitoring form
   - copies of the research lesson for the group participants

Observation Protocol (Selecting Roles for Observation)

1. Teaches lesson
   - Focus on teaching the research lesson as planned.
   - Select and sequence the discussion using the monitoring form.
2. Individual student(s)
   - Focus on interactions with the teacher and other students; write down what they say and do as it relates to the lesson; take pictures of their work (e.g., hands-on materials). The student is not a member of a group being observed or the observer responsible for the whole class and smaller groups.
   - Ask teacher of record for individual student(s) to observe such as a
     ◦ struggling student
     ◦ quiet student
     ◦ ELL student
     ◦ SPED student
     ◦ student who works but does not show work
3. Group interaction (focus group[s])
   - Focus on interactions among group members including conversations between students and between students and teacher; write down student conversations as they relate to the content; record any off-task behavior; take pictures of their work (e.g., hands-on materials).

- Select one group or if selecting two, sit between them so both can be observed from a seated position. The group does not include the individual student who is being observed or the observer responsible for the whole class and smaller groups.
4. Questions asked by the teacher
   - Focus on questions asked by the teacher; write down questions; take note of the position of the teacher in the room and also if directed of the whole class, small group, or individual.
   - Select a seat in the room that allows for taking note of whole-class comments and also of those among another group or two groups that are different from the group(s) selected by the observer of the group, the individual student, or the observer responsible for the whole class and smaller groups.
5. Questions/statements/explanations asked by the students
   - Focus on questions and statements asked by the student; write down questions; take note of student-to-student or student-to-teacher interactions; take note of sequenced student ideas based on the monitoring form.
   - Select a seat in the room that allows for taking note of whole-class comments and also of those among another group or two groups that are different from the group(s) selected by the observer of the group and also the individual student.

IMPLEMENTING THE RESEARCH LESSON

Teaching Guidelines for the Research Lesson

- As the teacher, you are responsible for collecting student work along the way (such as taking pictures of their materials with an iPad).
- Teach the lesson as planned, but if any modifications are made, make note of the reason for the changes.
- Complete the monitoring form as the lesson is taught.

Observing Guidelines for the Research Lesson

- As an observer, you will not be able to speak to the students or assist in *any way*. You are there to take notes on your observations. Do not interact.
- When recording exchanges among students, record the names of the students in your notes.
- Take note of visual aids (blackboard, overhead, etc.) and materials used.
- Have your iPad charged and ready for use.

DEBRIEFING THE RESEARCH LESSON

Debriefing Protocol

The focus of the debriefing session is to answer this question: "What do you really know about what the students know about the concept and how do you know it?"

- Teaches lesson as planned.
  - Describe how you believe lesson was executed.
  - Describe the use of the monitoring form.
  - Note what changes you would make.
  - Describe what you noticed about student learning.
- Individual student(s)
  - Describe how the student interacted with the teacher and other students.
  - Describe the student's nonverbal cues (e.g., body language, eye contact).
  - Describe the student's thinking related to the content being taught with evidence (e.g., pictures, student work).
  - What can you infer about the student's understanding of the concept?
- Group interaction (focus group[s])
  - Describe interactions between students and between students and teacher.
  - Describe any of the students' nonverbal cues (e.g., body language, eye contact).
  - Describe the students' thinking related to the content being taught with evidence (e.g., pictures, student work).
  - What can you infer about the students' understanding of the concept?
- Questions asked by the teacher
  - How did the questions facilitate student learning of the content? Give examples.
  - Describe the use of open- and closed-ended questions.
  - Describe the selection of the audience for the question (individual, small group, whole class).
  - Describe the use of wait time.
  - Describe the use of follow-up questioning to scaffold instruction.
- Questions/statements/explanations from the students
  - Describe the quality of the students' questions that demonstrates their attempts to understand the concept.
  - Describe students' questions and responses that reflect self-evaluation of their own learning.
  - Describe students' thinking in terms of the monitoring form.

REFERENCES

Anderson, R. A., Crabtree, B. F., Steele, D. J., & McDaniel, R. R. (2005). Case study research: The view from complexity science. *Qualitative Health Research, 15*(5), 669–685.

Aoki, T. (2005). Legitimating live curriculum: Toward a curricular landscape of multiplicity. In W. Pinar & R. Irwin (Eds.), *Curriculum in a new key: The collected works of Ted T. Aoki* (pp. 199–215). Mahwah, NJ: Lawrence Erlbaum Associates.

Avalos, B. (2011). Teacher professional development in *Teaching and Teacher Education* over ten years. *Teaching and Teacher Education, 27*(1), 10–20.

Bateson, G. (2002). *Mind and nature: A necessary unity.* New York, NY: Bantam Books. (Original work published 1979)

Black, P. (2004). Working inside the black box: Assessment for learning in the classroom. *Phi Delta Kappan, 86*(1), 9.

Black, P., & Wiliam, D. (1998). Assessment and classroom learning. *Assessment in Education: Principles, Policy & Practice, 5*(1), 7.

Blank, R. K., de las Alas, N., & Smith, C. (2007). *Analysis of the quality of professional development programs for mathematics and science teachers: Findings from a cross-state study.* Washington, DC: Council of Chief State School Officers.

Burbules, N. C., & Rice, S. (2010). On pretending to listen. *Teachers College Record, 112*(11), 2874–2888.

Bybee, R. W. (2014). The BSCS 5E instructional model: Personal reflections and contemporary implications. *Science and Children, 51*(8), 10–13.

Bybee, R. W., Taylor, J. A., Gardner, A., Van Scotter, P., Carlson Powell, J., Westbrook, A., & Landes, N. (2006). *BSCS 5E instructional model: Origins and effectiveness.* Colorado Springs, CO: BSCS.

Chinnappan, M., & Cheah, U. (2012). Mathematics knowledge for teaching: Evidence from lesson study. In J. Dindyal, L. P. Cheng, & S. F. Ng (Eds.), *Mathematics education: Expanding horizons. Proceedings of the 35th Annual Conference of the Mathematics Education Research Group of Australasia* (pp. 194–201). Singapore: MERGA.

Cochran-Smith, M., Ell, F., Grudnoff, L., Haigh, M., Hill, M., & Ludlow, L. (2016). Initial teacher education: What does it take to put equity at the center? *Teaching and Teacher Education, 57,* 67–78.

Cochran-Smith, M., Ell, F., Grudnoff, L., Ludlow, L., Haigh, M., & Hill, M. (2014a). When complexity theory meets critical realism: A platform for research on initial teacher education. *Teacher Education Quarterly, 41*(1), 105–122.

Cochran-Smith, M., Ell, F., Ludlow, L., Grudnoff, L., & Aitken, G. (2014b). The challenge and promise of complexity theory for teacher education research. *Teachers College Record, 116*(5), 1–38.

Davis, B. (1997). Listening for differences: An evolving conception of mathematics teaching. *Journal for Research in Mathematics Education, 28*(3), 355–376.

Davis, B., & Simmt, E. (2016). Perspectives on complex systems in mathematics learning. In L. English & D. Kirshner (Eds.), *Handbook of international research in mathematics education, 3rd ed.* (pp. 416–432). New York, NY: Routledge.

Davis, B., Sumara, D., & Luce-Kapler, R. (2000). *Engaging minds: Changing teaching in complex times.* New York, NY: Routledge.

Desimone, L. M. (2009). Improving impact studies of teachers' professional development: Toward better conceptualizations and measures. *Educational Researcher, 38*(3), 181–199.

Doll, W. E. (1993). *A post-modern perspective on curriculum.* New York, NY: Teachers College Press.

Doll, W. E. (2005). The culture of method. In W. Doll, J. Fleener, J. St. Julien, & D. Trueit (Eds.), *Chaos, complexity, curriculum, and culture: A conversation* (pp. 21–76). New York, NY: Peter Lang.

Doll, W. E., Fleener, M. J., St. Julien, J., & Trueit, D. (Eds.). (2005). *Chaos, complexity, curriculum, and culture: A conversation.* New York, NY: Peter Lang.

Doll, W. E., & Gough, N. (2002). *Curriculum visions.* New York, NY: Peter Lang.

Driscoll, M. (1999). *Fostering algebraic thinking: A guide for teachers, grades 6–10.* Portsmouth, NH: Heinemann.

Duckor, B. (2014). Formative assessment in seven good moves. *Educational Leadership, 71*(6), 28–32.

Eddy, C. M., & Harrell, P. E. (2013). *AssessToday: A short-cycle formative assessment observation protocol.*

Eddy, C. M., Harrell, P., & Heitz, L. (2017). An observation protocol of short-cycle formative assessment in the mathematics classroom. *Investigations in Mathematics Learning, 9*(3), 130–147. Retrieved from www.tandfonline.com/doi/full/10.1080/19477503.2017.1308699

Eddy, C. M., Quebec Fuentes, S., Ward, E. K., Parker, Y. A., Cooper, S., Jasper, B., Mallam, W., Sorto, M. A., & Wilkerson, T. (2015). Unifying the algebra for all movement. *Journal of Advanced Academics, 26*(1), 59–92.

Empson, S., & Jacobs, V. (2008). Learning to listen to children's mathematics. In D. Tirosh & T. Wood (Eds.), *International handbook of mathematics teacher education, vol. II: Tools and processes in mathematics teacher education* (pp. 257–281). Rotterdam, The Netherlands: Sense Publishers.

Fennell, F., Kobett, B. M., & Wray, J. A. (2017). *The formative 5: Everyday assessment techniques for every math classroom.* Thousand Oaks, CA: Corwin Mathematics/National Council of Teachers of Mathematics.

Hattie, J. (2008). *Visible learning: A synthesis of over 800 meta-analyses relating to achievement.* New York, NY: Routledge.

Heitz, L. (2013). *The validation of a short-cycle formative assessment observation protocol for science and mathematics instruction* (Unpublished doctoral dissertation). University of North Texas, Denton, TX. Retrieved from http://homepages.se.edu/lheitz/files/2013/12/Heitz-Dissertation-October-2013.pdf

Hirsh, S. (2009). A new definition. *Journal of Staff Development, 30*(4), 10–16.

Huebner, D. (1999a). An educator's perspective on language about God. In V. Hillis (Ed.), *The lure of the transcendent: Collected essays by Dwayne E. Huebner* (pp. 257–284). Mahwah, NJ: Lawrence Erlbaum Associates. (Original work published 1977)

Huebner, D. (1999b). New modes of man's relationship to man. In V. Hillis (Ed.), *The lure of the transcendent: Collected essays by Dwayne E. Huebner* (pp. 74–93). Mahwah, NJ: Lawrence Erlbaum Associates. (Original work published 1963)

Kadroon, T., & Inprasitha, M. (2013). Professional development of mathematics teachers with lesson study and open approach: The process for changing teachers' values about teaching mathematics. *Psychology, 4*(2), 101–105.

Kauffman, S. (1996). *At home in the universe: The search for the laws of self-organization and complexity.* New York, NY: Oxford University Press.

Koellner, K., & Jacobs, J. (2015). Distinguishing models of professional development: The case of an adaptive model's impact on teachers' knowledge, instruction, and student achievement. *Journal of Teacher Education, 66*(1), 51–67.

Lewis, C. C., & Perry, R. R. (2015). A randomized trial of lesson study with mathematical resource kits: Analysis of impact on teachers' beliefs and learning community. In J. Middleton, J. Cai, & S. Hwang (Eds.), *Large-scale studies in mathematics education* (pp. 133–158). Cham, Switzerland: Springer International.

Lewis, C., & Perry, R. (2017). Lesson study to scale up research-based knowledge: A randomized, controlled trial of fractions learning. *Journal for Research in Mathematics Education, 48*(3), 261–299. doi:10.5951/jresemathuc.48.3.0261

Lobato, J., Clarke, D., & Ellis, A. B. (2005). Initiating and eliciting in teaching: A reformulation of telling. *Journal for Research in Mathematics Education, 36*(2), 101–136.

Maskiewicz, A. C., & Winters, V. A. (2012). Understanding the co-construction of inquiry practices: A case study of a responsive teaching environment. *Journal of Research in Science Teaching, 49*(4), 429–464.

Mason, M. (Ed.). (2008). *Complexity theory and the philosophy of education.* Malden, MA: Wiley-Blackwell.

Maturana, H. R., & Varela, F. J. (1987). *The tree of knowledge: The biological roots of human understanding.* Boston, MA: New Science Library/Shambhala Publications.

Morin, E. (2008). *On complexity: Advances in systems theory, complexity, and the human sciences.* New York, NY: Hampton Press.

Pratt, S. (2008a). Complex constructivism: Rethinking the power dynamics of "understanding." *Journal of Canadian Association for Curriculum Studies, 6*(1), 113–132.

Pratt, S. (2008b). Bifurcations are not always exclusive. *Complicity, 5*(1), 125–128.

Pratt, S. (2018). The pedagogical complexity of story. In M. Quinn (Ed.), *Complexifying Curriculum Studies: Reflections on the Generative and Generous Gifts of William E. Doll, Jr.* New York, NY: Routledge.

Rodriguez, M. C. (2004). The role of classroom assessment in student performance on TIMSS. *Applied Measurement in Education, 17*(1), 1–24.

Smith, M. S., & Stein, M. K. (2011). *Five practices for orchestrating productive mathematics discussions.* Reston, VA: National Council of Teachers of Mathematics.

Stein, M., Engle, R., Smith, M., & Hughes, E. (2008). Orchestrating productive mathematical discussions: Five practices for helping teachers move beyond show and tell. *Mathematical Thinking and Learning, 10*, 313–340.

Trueit, D. (Ed.). (2012). *Pragmatism, post-modernism, and complexity theory: The "fascinating imaginative realm" of William E. Doll, Jr.* New York, NY: Routledge.

Wiliam, D., & Thompson M. (2007). Integrating assessment with instruction: What will it take to make it work? In C. A. Dwyer (Ed.), *The future of assessment: Shaping teaching and learning* (pp. 53–82). Mahwah, NJ: Lawrence Erlbaum Associates.

Chapter 10

Epilogue: Looking Toward the Future of Mathematics Teacher Preparation

Patrick M. Jenlink

Pre-service teachers preparing to enter classrooms in schools to teach mathematics at all grade levels need mathematics courses that develop a deep understanding of the mathematics they will teach. The mathematical knowledge needed by teachers at all levels is substantial. The "mathematical knowledge for teaching" needed by teachers is equally substantial. This knowledge for teaching mathematics is the knowledge used in recognizing, understanding, and responding to the mathematical problems and tasks encountered in teaching the subject of mathematics today (Ball & Hill, 2008; Ball, Thames, & Phelps, 2008; Copur-Gencturk, 2015; Kleickmann et al., 2013; Phelps & Howell, 2016).

Pre-service teachers need to understand "the intricate combinations of critical component skills such as concepts, procedures, algorithms, computation, problem solving, and language" (Riccomini, Smith, Hughes, & Fries, 2015, p. 236) that underlie mathematics as it is intended to be taught so that they can teach it to diverse groups of students as a coherent, reasoned activity and communicate an appreciation of the elegance and power of the subject (Hill et al., 2008; Riccomini, Sanders, & Jones, 2008). With such language and knowledge, teachers can foster an enthusiasm for mathematics and a deeper understanding among their students.

College courses necessary for developing this mathematical language and knowledge should make connections between the mathematics being studied and the mathematics pre-service teachers will teach. Determining the "mathematics content knowledge" as well as "mathematics knowledge for teaching" requires bridging between content area specialists in mathematics and teacher preparation faculty. In making the determination, it is important to ensure that preparing teachers to teach mathematics in school classrooms focuses on not only the content knowledge of mathematics but equally on using this knowledge in teaching practice.

The caution here is to avoid significant gaps in mathematics knowledge in students as they matriculate through grade levels and eventually move to the college classroom (Copur-Gencturk & Lubienski, 2013). Advanced mathematical knowledge, for the teacher, is necessary for advanced mathematical thinking and logical reasoning in preparing lessons and engaging in pedagogical practices for teaching (Zazkis & Leikin, 2010).

Preparing teachers to teach mathematics, specifically the mathematical education of teachers, should be seen as collaboration between mathematics faculty who are content specialists responsible for theoretical knowledge and logical reasoning and teacher preparation faculty responsible for pedagogical content knowledge for mathematics education (Hauk, Toney, Jackson, Nair, & Y Tsay, 2014; Hill, Ball, & Schilling, 2008; Hill et al., 2008).

High-quality mathematics instruction in classrooms and schools, at all grade levels, involves a combination of mathematical knowledge and pedagogy such as choosing appropriate examples and teaching strategies for various topics. Equally important is the "mathematical knowledge for teaching" that merges the language of mathematics and the substantive content of mathematics with the pedagogical skills required to be a quality teacher (Steele & Rogers, 2012; Suurtamm & Vezina, 2010). Also important is pedagogical content knowledge, which is an inextricable blending that is "predicated on coherent and generative understandings of the big mathematical ideas that make up the curriculum" (Silverman & Thompson, 2008, p. 502). In this sense, pedagogical content knowledge is more than an overlap of knowledge that is both pedagogically and mathematically connected.

Importantly, the collaboration between teacher preparation faculty and mathematics faculty as content specialists ensures a co-joined responsibility for preparing mathematics teachers. Teacher education faculty can provide valuable insights and information about what takes place in school classrooms, including common mathematical misunderstandings of practicing teachers and how to build on these to promote real understanding.

Whereas teacher preparation faculty have access to information on state curriculum guidelines and research studies about teachers' mathematical knowledge, mathematics faculty have insight as to the language and knowledge of mathematics that can inform logical reasoning and pedagogical practices essential to understanding mathematical knowledge for teaching. Mathematics faculty as content area specialists can help mathematics education faculty by keeping them informed of mathematical developments that may have an impact on mathematics for the school classroom.

The inescapable realization is that a powerful relationship exists between what a teacher knows, how he or she knows it, and what he or she can do in the context of designing and delivering instruction in the classroom (Hill et al., 2008; Phelps & Howell, 2016). What a teacher knows in terms of mathematics content knowledge and mathematical knowledge for teaching and pedagogical content knowledge is predicated on the sophistication, quality, and design of the mathematics and teacher preparation courses he or she experienced in preparing to be a teacher and enter a classroom to teach mathematics.

FINAL REFLECTIONS

Anthony and Walshaw (2009), writing on the characteristics of effective mathematics teaching, noted that it is widely understood mathematics plays a key role in shaping how individuals deal with the various facets of their lives. Few would argue otherwise. Mathematics as a universal language is a critical part of how we function, whether consciously or unconsciously. Knowing the importance of mathematics in the lives of individuals and how those lives, based on an understanding of mathematics, are shaped and at the same time shape the world around them, we have a responsibility to ensure that students in classrooms studying mathematics receive the highest quality of instruction possible.

That said, we must understand what effective mathematics teaching looks like (Anthony & Walshaw, 2009). To that end, we have a responsibility to prepare mathematics teachers with the language and knowledge of mathematics. Teachers need the opportunity to develop their understanding of mathematics language and their knowledge of how to teach mathematics. This focuses responsibility on teacher preparation programs and the necessity to form collaborations with mathematics faculty to ensure that teachers receive the very best preparation experiences possible.

REFERENCES

Anthony, G., & Walshaw, M. (2009). Characteristics of effective mathematics: A view from the West. *Journal of Mathematics Education, 2*(2), 147–164.

Ball, D. L., & Hill, H. C. (2008). *Mathematical knowledge for teaching (MKT) measures: Mathematics released items 2008.* http://sitemaker.umich.edu/lmt/files /LMT_sample_ items.pdf

Ball, D. L., Thames, M. H., & Phelps, G. (2008). Content knowledge for teaching: What makes it special? *Journal of Teacher Education, 59*(5), 389–407. doi:10.1177/0022487108324554

Copur-Gencturk, Y. (2015). The effects of changes in mathematical knowledge on teaching: A longitudinal study of teachers' knowledge and instruction. *Journal for Research in Mathematics Education, 46*(3), 280–330.

Copur-Gencturk, Y., & Lubienski, S. T. (2013). Measuring mathematical knowledge for teaching: A longitudinal study using two measures. *Journal of Mathematics Teacher Education, 16*(3), 211–236.

Hauk, S., Toney, A., Jackson, B., Nair, R., & Y Tsay, J-J. (2014). Developing a model of pedagogical content knowledge for secondary and post-secondary mathematics. *Dialogic Pedagogy, 2*, 16–40. http://dpj.pitt.edu

Hill, H. C., Ball, D. L., & Schilling, S. G. (2008). Unpacking pedagogical content knowledge: Conceptualizing and measuring teachers' topic-specific knowledge of students. *Journal for Research in Mathematics Education, 39*(4), 372–400.

Hill, H. C., Blunk M. L., Charalambous, C. Y., Lewis, J. M., Phelps, G. C., Sleep, L., & Ball, D. L. (2008). Mathematical knowledge for teaching and the mathematical quality of instruction: An exploratory study. *Cognition and Instruction, 26*(4), 430–511.

Kleickmann, T., Richter, D., Kunter, M., Elsner, J., Besser, M., Krauss, S., & Baumert, J. (2013). Teachers' content knowledge and pedagogical content knowledge: The role of structural differences in teacher education. *Journal of Teacher Education, 64*(1), 90–106.

Phelps, G., & Howell, H. (2016). Assessing mathematical knowledge for teaching: The role of teaching context. *Mathematics Enthusiast, 13*(1–2), 52–70.

Riccomini, P. J., Sanders, S., & Jones, J. (2008). The key to enhancing students' mathematical vocabulary knowledge. *Journal on School Educational Technology, 4*(1), 1–7.

Riccomini, P. J., Smith, G. W., Hughes, E. M., & Fries, K. F. (2015). The language of mathematics: The importance of teaching and learning mathematical vocabulary. *Reading & Writing Quarterly, 31*(3), 235–252. doi:10.1080/10573569.2015.1030995

Silverman, J., & Thompson, P. W. (2008). Toward a framework for the development of mathematical knowledge for teaching. *Journal of Mathematics Teacher Education, 11*, 499–511.

Steele, M. D., & Rogers, K. C. (2012). Relationships between mathematical knowledge for teaching and teaching practice: The case of proof. *Journal of Math Teacher Education, 15*, 159–180.

Suurtamm, C., & Vezina, N. (2010). Transforming pedagogical practice in mathematics: Moving from telling to listening. *International Journal for Mathematics Teaching and Learning, 31*, 1–19. Retrieved from www.cimt.plymouth.ac.uk/journal/default.htm

Zazkis, R., & Leikin, R. (2010). Advanced mathematical knowledge in teaching practice: Perceptions of secondary mathematics teachers. *Mathematical Thinking and Learning, 12*(4), 263–281.

About the Editor and Contributors

EDITOR

Patrick M. Jenlink is regents professor in the College of Education, Stephen F. Austin State University. He has served as E. J. Campbell Endowed Chair in Educational Leadership, doctoral program coordinator, and department chair during his tenure. Dr. Jenlink's teaching emphasis in doctoral studies includes courses on ethics and philosophy of leadership, research methods and design, and leadership theory and practice. Dr. Jenlink's research interests include politics of identity, democratic education, self-efficacy theory, and critical theory. He has edited or authored over 12 books and authored over 70 book chapters. He has also authored and published over 175 peer-refereed articles and over 200 peer-refereed conference papers. His most recent books include *STEM Teaching: An Interdisciplinary Approach* (Rowman & Littlefield); *Teacher Preparation at the Intersection of Race and Poverty in Today's Schools* (Rowman & Littlefield); *Multimedia Learning Theory and Its Implications for Teaching and Learning* (Rowman & Littlefield); and the *Dewey Studies Handbook* (Brill/Sense). His current book projects include *Fostering Teachers' Sense of Efficacy and Commitment to Teaching: Toward More Efficacious Teacher Preparation* (Rowman & Littlefield); and *Ethics and the Educational Leader: A Casebook of Ethical Dilemmas* (Rowman & Littlefield).

CONTRIBUTORS

Didem Akyuz is an associate professor in the Department of Mathematics and Science Education at Middle East Technical University. Her studies focus on technology in mathematics education; teacher practices, beliefs, and goals; and mathematics teacher education. Dr. Akyuz also gives lectures to graduate and undergraduate elementary mathematics education students and conducts research based on the practices of teachers and teacher candidates.

Erdinc Cakiroglu is a professor in the Department of Mathematics and Science Education at Middle East Technical University. His current work involves the use of technology in mathematical problem solving; mathematics teachers' professional development; curriculum development in mathematics and representations of mathematics concepts; self-regulation of learning to teach; teachers' use of curriculum materials; and learning numeracy at the elementary level. Dr. Cakiroglu has recently published a book chapter related to the defining process of mathematical concepts.

Colleen McLean Eddy is an associate professor in the Department of Teacher Education and Administration at the University of North Texas, where she joined the faculty in 2006. Her scholarship focuses on teacher preparation and teacher quality in mathematics education. Her research has been supported by grants totaling over $5 million, including the National Science Foundation Robert Noyce Scholarship Grant for which she is the principal investigator, and a Teacher Quality Partnership Grant Program grant leading to nine years as the project director for XTreem Math. At the University of North Texas she teaches graduate and undergraduate courses in mathematics education, curriculum and instruction, and secondary education.

Mindy Eichhorn is an assistant professor in the education department at Gordon College. Mindy currently teaches courses on special education assessment, the Individualized Education Program process, and inclusion while supervising student teachers and math specialist degree candidates. Her research interests include math learning disabilities, transition, early intervention in mathematics, teachers' perceptions of mathematics, and the use of professional development to improve math instruction. She is also a mathematics specialist for the Boston Children's Hospital Learning Disabilities Program. She presently serves as unit planner for the Global Mathematics Special Interest Group with the Comparative and International Education Society.

Helen Elizabeth Garnier is a consultant for the Institute of Cognitive Science, University of Colorado, Boulder, and she collaborates on research aimed at supporting teacher and student mathematics, science, and literacy learning. Dr. Garnier developed two international databases for the TIMSS 1999 Video Study of Mathematics and Science and created the accompanying documentation. She collaborated with Dr. Jessaca Spybrook on a three-level mediation model to identify the mechanisms by which a school-based professional development program predicted improved student achievement through quality of classroom instruction. She also collaborated with Dr. Lindsay Clare Matsumura to develop, refine, and provide technical information on the Instructional Quality Assessment. She has published the *Third International Mathematics and Science Study 1999 Video Study Technical Report, Volume 2: Science* (NCES 2011-049).

Tracy Goodson-Espy is a professor in the Department of Curriculum and Instruction at Appalachian State University. Tracy is interested in STEM teacher education at elementary and secondary levels. She is also assistant director of the doctoral program at the university and is responsible for teaching quantitative and qualitative research courses in that program.

Michelle C. Hughes, EdD, is an associate professor of education at Westmont College, Santa Barbara, California. Michelle teaches and supervises undergraduates earning a bachelor's degree in liberal studies with elementary or secondary teaching credentials. Michelle is a former junior high English teacher and high school administrator.

Jennifer Jacobs is an associate research professor at the University of Colorado, Boulder, in the Institute of Cognitive Science. Dr. Jacobs's research focuses on the classroom teaching and learning of mathematics and science and models of professional development to support teachers and their students. Dr. Jacobs is principal investigator for the Learning and Teaching Geometry Efficacy Project and a co–principal investigator for SchoolWide Labs—a real-time sensing and data logging platform for integrating computational thinking into middle school STEM curricula and the TalkBack application. She has co-authored *Learning and Teaching Geometry: Video Cases for Mathematics Professional Development, Grades 6–12*; and *Mathematics Professional Development: Improving Teaching Using the Problem-Solving Cycle and Leadership Preparation Models*.

Karen Koellner, a professor at Hunter College, City University of New York, has been designing and researching mathematics professional development models for the past 20 years. These models focus on supporting secondary teachers to deeply learn mathematical content and practices with the goal of providing access and high-quality instruction to students in all contexts. Dr. Koellner is currently principal investigator for the National Science Foundation–funded study Taking a Deep Dive: Investigating PD Impact on What Teachers Take Up and Use in Their Classrooms and a co–principal investigator for the Learning and Teaching Geometry Efficacy Project. Dr. Koellner has co-authored the book *Mathematics Professional Development: Improving Teaching Using the Problem-Solving Cycle and Leadership Preparation Models.*

Carmen M. Latterell is a professor in the Department of Mathematics and Statistics at the University of Minnesota, Duluth, where she conducts research in mathematics education. Currently her research is taking a philosophical direction as she tries to delineate how various groups of people view mathematics. She has published two books: *Math Wars*; and *Helping Students Prepare for College Mathematics Placement Tests.*

Diana L. Moss is an assistant professor in mathematics education in the Department of Curriculum and Instruction at Appalachian State University. She teaches junior and senior undergraduate courses in mathematics content and pedagogy and is a member of the graduate faculty. Her teaching and research interests include children's mathematical thinking and teacher education. Her work involves using design research in classroom settings to support learning through the perspective of social constructivism.

David Nurenberg is core faculty in the middle and high school certification program at Lesley University Graduate School of Education where he teaches courses in lesson planning, curriculum design, classroom management, and project-based learning. His research involves student-centered pedagogies, and his work has been published several times in peer-reviewed journals. Dr. Nurenberg consults with schools in the Boston area on matters of student-centered pedagogy.

Nanette Seago, a senior research associate at WestEd STEM, has been designing and researching mathematics professional development materials for the past 20 years to prepare middle and high school mathematics teachers to more effectively teach challenging mathematical concepts. Ms. Seago is currently principal investigator for the Video in the Middle Project, co–principal investigator for the Learning and Teaching Geometry Efficacy Project, and co–principal investigator for Taking a Deep Dive: Investigating PD Impact on What Teachers Take Up and Use in Their Classrooms. She has designed, developed, and published two personal development materials: *Learning and Teaching Linear Functions: Video Cases for Mathematics Professional Development*; and *Teaching Geometry: Video Cases for Mathematics Professional Development*. In addition, she has published the book *Examining Mathematics Practice through Classroom Artifacts*.

Se-Ah Kwon Siegel is the associate director of assessment at Lesley University Graduate School of Education. Ms. Siegel designs data sets and assessment methods to measure progress in student learning across the school's educational programs and is responsible for producing reports for state and Council for the Accreditation of Educator Preparation accreditation. Her research interests include assessment models that capture student learning and models that measure progress for improvement efforts.

Sinem Sozen Ozdogan is a research assistant in the Faculty of Education at TED University in Ankara. She is also a PhD candidate in the Faculty of Education in the Department of Elementary Education at Middle East Technical University. Her research focuses on the history of mathematics in the mathematics education field.

Sarah Smitherman Pratt is an assistant professor in the Department of Teacher Education and Administration at the University of North Texas. Her research and teaching focus on the intersection of mathematics education and curriculum theory, and her special interest is in complexity theories as they relate to complex conversations in education. She has served as chair and program chair for AERA SIG: Chaos and Complexity Theories as well as secretary and member coordinator for the Research Council on Mathematics Learning. She teaches undergraduate and graduate courses that focus on learning theories, mathematics education, and curriculum theory.

Lisa L. Poling is an associate professor in the Department of Curriculum and Instruction at Appalachian State University. She currently teaches junior- and senior-level undergraduate courses in mathematic pedagogy and content and is a member of the graduate faculty. Her research interests focus on the teacher's sense of responsibility in educating all children in the science of mathematics and on utilizing mathematics to critically assess social issues.

Courtney Vitale is an undergraduate student at Gordon College studying elementary education and mathematics with a concentration in special education. Courtney is interested in research pertaining to pre-service teachers' perceptions surrounding mathematics, engaging students in mathematics, and math phobia and stereotypes.

Chao Wang is a research associate at the School of Education, Institute of Cognitive Science, University of Colorado, Boulder, and specializes in linguistic and sociocultural issues that are relevant to the academic success of culturally and linguistically diverse students. Ms. Wang's research work focuses on methods for improving instructional and assessment practices for English language learners in science and mathematics. She has taught three graduate-level courses: Diagnostic Testing in Bilingual Education; Research and Evaluation in Bilingual Education; and Parent and Community Involvement.

Janelle L. Wilson is a professor in the Department of Anthropology, Sociology & Criminology at the University of Minnesota, Duluth, where she teaches courses in social psychology and deviance. Her research interests include the sociology of everyday life, nostalgia, and generational identity. She is author of the book *Nostalgia: Sanctuary of Meaning* (Bucknell University Press).